THE FINGER OF GOD

To Ruth

THE FINGER OF GOD

How God's Existence Impacts Upon Humanity

VINCENT JOHN DELANY

BRIGHTON • PORTLAND

Copyright © Vincent John Delany, 2009

The right of Vincent John Delany to be identified as Author of this work has been asserted in accordance with the Copyright, Designs and Patents Act 1988.

2 4 6 8 10 9 7 5 3

First published 2009 in Great Britain by
THE ALPHA PRESS
PO Box 139
Eastbourne BN24 9BP

and in the United States of America by
THE ALPHA PRESS
920 NE 58th Ave   Suite 300
Portland, Oregon 97213-3786

All rights reserved. Except for the quotation of short passages for the purposes ofcriticism and review, no part of this publication may be reproduced, stored in a retrieval system, or transmitted, in any form or by any means, electronic, mechanical, photocopying, recording or otherwise, without the prior permission of the publisher.

British Library Cataloguing in Publication Data
A CIP catalogue record for this book is available from the British Library.

ISBN 978-1-898595-54-0

Scripture quotations taken from The Holy Bible, New International Version Anglicised Copyright © 1979, 1984 by International Bible Society. Used by permission of Hodder & Stoughton Publishers, A division of Hachette Livre (UK) Ltd. All rights reserved. "NIV" is a registered trademark of International Bible Society. UK trademark number 1448790. Scripture quotations also taken from the Authorised Version.

Typeset and designed by THE ALPHA PRESS, Brighton & Eastbourne.
Printed by TJ International, Padstow, Cornwall.
This book is printed on acid-free paper.

Contents

Preface vii
Acknowledgements xi

Introduction 1

Part One God's Present Kingdom

 1 The Nature of the Kingdom 11
 2 History of the Kingdom 16
 3 The Hidden Kingdom 22
 4 Hidden from Citizens 32
 5 Entering the Kingdom 38
 6 Life in the Kingdom 41
 7 Faith and the Kingdom 46
 8 Servants of the Kingdom 53
 9 The Kingdom and the Word 58
10 The Holy Spirit and the Kingdom of God 63
11 Evil and the Kingdom 70
12 Horoscopes 75
13 Hidden from the Wise 80
14 Testimonies of the Kingdom 86
15 Prayer and the Kingdom 92
16 Denouement 96

Part Two The Coming Kingdom

17 Alternative Eschatology 107
18 End Days I: The Olivet Discourse 114
19 End Days II: Revelation of John the Apostle 124
20 The Fall of Babylon 134
21 Christ's Kingdom 140
22 The New Jerusalem 148

Part Three Summary and Conclusions

23 A Review: Mystery, Majesty, Immanence   157

Bibliography   171
About the Author   173
Bible Reference Index   174
Subject Index   187

Preface

> "Wherefore, we receiving a kingdom which cannot be moved, let us have grace whereby we may serve God acceptably with reverence and godly fear: for our God is a consuming fire." *Hebrews 12:28, 29*

For many years I was not taken with the sound of bagpipes. The initial droning and following discordance offended my ears. However, I only ever heard them play when as a boy I listened on the family gramophone or on my father's accumulator activated wireless set (much to his annoyance). He had been unfortunate enough to be introduced to this windy cacophony as a prospective piper. Disastrously, practice was conducted with six other learners in a block of flats in North London. They marched around a dining room table. The noise was even weirder than normal on account of the rawness of their efforts. Putting it mildly, they were making uncertain notes! He only shared this experience in old age. The combined effect of the deafening noise and the subsequent hostility of the other occupants of the block caused him to either erase the memory from his mind or decide to seal his lips. It was not until my wife and I were on a holiday in the Ardnamurchan Peninsular of the Scottish Highlands that I underwent a change of attitude. From the shores of a loch the strains of pipes wafted softly over the waters from a very far distance. The sound was beautiful and very moving. There was, I thought, a lot more to the pipes than I had ever previously imagined. Not only were the strains melodious in the environ of the glens but their inventor obviously had a deep sense of purpose. That was as a clarion call. Well did Sir Walter Scott pen vividly:

Pibroch of Donuil Dhu
Wake thy wild voice anew,
Summon Clan Conuil.
Come away, come away,
Hark to the summons!
Leave untended the herd,
The flock without shelter;
Leave the corpse uninterr'd,

The bride at the altar;
Leave the deer, leave the steer,
Leave nets and barges:
Come as the winds come , when
Forests are rended,
Comes as the waves come, when
Navies are stranded:
Fast they come, fast they come;
See how they gather.
Wide waves the eagle plume
Blended with heather.

Clansmen's very lives depended upon the effectiveness of this summons. Only in the light of such an overriding emergency could it possibly have been acceptable and rational to abandon so many otherwise essential activities. Even the poet's eloquence hardly conveys the full degree of fear and sense of urgency that the sound instilled in the clansmen. Indeed, it is difficult for anybody to understand the trauma unless one has memories of hearing an air-raid siren in wartime.

A close friend, born in Czechoslovakia, was given options. He could either join the German army or be shot. He was one of seven survivors of a whole battalion who were taken prisoner. He gladly opted to join the Free Polish Army under an assumed name. Whenever he hears bagpipes his stomach churns over. The sounds used to come to him frighteningly and eerily in moonlight just before an opposing highland division made attacks. The notes were hostile in his ears. Forty years later he heard and responded to the "sweet sound" of amazing grace!

There was an alarming and more compelling summons which came over two millennia ago. A seemingly wild, ascetic, and frightening man emerged from the Judean Desert. He was living on locusts and wild honey and wore clothes made out of camel hair. This was a voice crying in the wilderness. It was that of John the Baptist. He cried, "Repent for the Kingdom of Heaven is at Hand!" He came to witness that the true light that gives light to every person was coming into the world. Hearers were drawn by the finger of God. His message went further. It heralded the coming of One who was the only begotten Son of God. Grace and truth would come by him. Although John baptised with water, he would baptise with the Holy Spirit. He would usher in the kingdom as a present reality. Not only should people repent (change) but also show evidence of this in their lives. Of this One, John was not worthy to unlatch a thong of his sandal. Jesus was this person. John's message needed to be unfolded and enlarged by the true light that lights everyone.

I cannot play the pipes nor do I have the charisma of John the Baptist. In any case the clarion call of pipes by comparison with these Galilean events is like comparing dim candlelight with the incandescence of a halogen bulb. But the poet's canny sense of urgency does help us to bring into perspective the teaching of Jesus that unless disciples were willing to leave father and mother and follow him they were not worthy of him. The very voice of Jesus is what must stir us today. For a time is coming when all who are in their graves will hear his voice and come out – those who have done good will rise to live, and those who have done evil will rise to be condemned. My burden is to urge people to go back to the certain sound of the original message of Jesus, loud and clear so that it drowns out the discordant and contradictory messages of Christendom. Of these it must be asked that if the trumpet makes an uncertain note who will prepare himself for the battle? A "certain note" is not only an audible sound but a call to a life that is to be lived. The evidence of having passed from darkness to light is love for people. Without love one is still spiritually dead. Lack of love is what is endemically wrong with the present world.

The opening scripture shows categorically that believers receive the kingdom here and are to serve within it now. This is supported by many other scriptures. For example, the Apostle Paul wrote to the Colossian church, "For he (the Father) has rescued us from the dominion of darkness and brought us into the kingdom of his son." Commentators also support this teaching but it has not gripped and motivated enough people to achieve God's optimum objectives. This is why in my old age I have written this book in an attempt to persuade many more to embrace this truth. It would then follow, prospectively, that the elusive goal of Christian unity could be achieved, and that passive believers could be motivated to join the comparatively few who already labour valiantly in God's vineyard.

It is, however, doubtful whether these hopes will be fulfilled to any extent during the ongoing history of this present world. It will be sufficient at this juncture to leave two thoughts. The alarming summons of John the Baptist, who was a man who came from God, set in motion a course of events such as the world had never experienced or could have imagined. It was a momentous and epoch shattering precursor of the beginning of God's great rescue operation. The views set out in Part Two, concerning God's ultimate triumph, are ground-breaking. These cover His mighty final great victory over darkness and the establishment of His eternal kingdom.

In dealing with the history of the visible Christian church (as distinct from the kingdom) I found it necessary to include strictures. I tried to do so politely and without bias against any one denomination. I trust I have shown an understanding of individuals' motives. Readers will react according to how literally they regard the message of Jesus. I ask non-believers and agnostics to exercise patience with what they perceive as theological jargon as I would not want them to miss many real-life accounts scattered throughout that should be of considerable interest to them.

While all scriptural references are given in full, the purpose of this book and its intended audience preclude the necessity of providing detailed sources for more contemporary events and personages, which should be understandable and recognized from the context in which this information is provided.

In that there is merit in this work it is due to the One who can guide and help. In that there are deficiencies these are due to the fallibility of the author. Those who have been touched by the Finger of God have him as their strength in this life and as a blessed hope for the future. Those who have not are spiritually alone, being without God and devoid of any tangible hope.

The views expressed are entirely those of the author and not necessarily of any other person who has helped in its production.

Acknowledgements

Having determined to be "my own man" and to resort only to the Bible and draw on what was in my memory I thought that my only debt to other authors was what I may sub-consciously have gained from their writings over the years. I have therefore included a short Bibliography. Also, having been presented with Professor Bauckham's scholarly work, *A Climax of Prophecy*, by Dame Janet Trotter, I became indebted to him for tracing a number of Exodus scripture links to The Song of Moses. My optimism about the ultimate victory of God over evil and the extent to which the nations will be won was also underpinned by his exposition.

The late Dr. Martyn Lloyd Jones is mentioned in chapter 14. I record here my gratitude to him for reading my printed lecture notes, endorsing them and recommending these should be expanded and published as a book.

I am grateful to Rebecca Burnett and young Daisy Marlowe for their untiring support in word processing, especially in helping to construct the index. I must also pay tribute to Ruth, my wife of 55 years, a former voluntary worker with deaf/blind persons, and a visitor to hospitals and residential homes until cruelly laid aside in 2005 as a result of contracting viral pneumonia, and for whom I now care. She has cheerfully allowed me to spend time on writing this book.

"The Lord answered Job out of the whirlwind and said . . . where wast thou . . . when the morning stars sang together, and all the sons of God shouted for joy? . . Canst thou bind the sweet influences of Pleiades, or loose the bands of Orion?"

Job 38:1, 4, 7, 31
Authorised Version

Introduction

"If I by the finger of God cast out demons, then the kingdom of God has come upon you." *Luke 11:20*

Over the two thousand years since Jesus rose from the dead faith in him has grown and proliferated to a remarkable extent. So has his teaching about the Father God he came to glorify. He foreshadowed this growth in a story about a small grain of mustard seed that grows into a large tree, so that the birds of the air come and perch in its branches. It is instructive that Jesus applied this illustration to the progress of "the kingdom of God" thus showing that God's everlasting kingdom was in the world working actively. He was careful not to apply this parable to the visibly organised church. "The birds of the air" obviously referred to factors that were intrusive and superfluous. Jesus used this latter illustration on another occasion to show how extraneous factors impede the spread of the word. In the parable of the sower the birds of the air eat the seed before it can take root. The overall lesson to be learned is that the kingdom of God is of self-standing importance, and that its propagation should not suffer detraction or embellishment. It will suffer opposition and it will suffer religious competition. Its motivation is derived solely from the power of the Father or in other words by "the finger of God." As evidence one has only to look at the predominant teaching in the early ministry of Jesus and his followers, and the power that accompanied it; it was about the kingdom of God. It was the first thing Jesus talked about after he rose from the dead, and he revealed much more about its future to his beloved John in visions on the Isle of Patmos.

The kingdom of God will stand unchanging. The gates of hell will not prevail against it. Over the centuries it has boasted stalwart workers (or citizens) who endured persecution, derision, hardships, and even martyrdom. This must not be lost sight of when the failings of the visible church are discussed. In Part One, positive and uplifting aspects of the true kingdom will be set before readers. Namely, the nature of the kingdom (chapter 1), its history (chapter 2), how it is entered (chapter 5), its life (chapter 6), its faith (chapter 7), its servants, (chapter 8) its word (chapter 9), its power (chapter 10), its testimonies (chapter 14) and its prayer (chapter 15). Then there are explanations of how the kingdom can be hidden from different people ranging from church members, non-church going believers, agnostics, and those members of the public at large who are ambivalent about spirituality and life after death. Knowledge is impeded in various ways according to peoples' spiritual understanding or their lack of it. This is discussed in two separate chapters (3 & 4) in which people are classified as "citizens" (of the kingdom) or "strangers" (from it). No disrespect is intended by the adage "stranger". A centurion, a Syrophenician woman, and a good Samaritan, were not ostensibly formerly believers but their faith earned the admiration of Jesus. He showed no antipathy. In ancient Israel strangers were not to be oppressed. Indeed crops were to be left in the corners of fields for their benefit. There was a place at the Passover table for them. The Jews had learned what it meant to be strangers in Egypt. This is a salutary lesson because visible churches throughout the ages were seldom as accommodating and had not learned anything from earlier persecutions of believers firstly by the Sanhedrin and then Roman emperors. This is surely also a lesson for a number of other faiths at the present day in various parts of the world.

Just as strangers are to be respected so must the right of personal freedom. In order to put any implied criticism of Christendom or individuals in perspective three caveats are entered. Firstly, it would be a sad error to engage in unjustified scandal mongering. Hypocritical attitudes only add fuel to the fires of disaffection and delight atheists. Therefore it is not sought here to condemn every aspect of denominationalism or to offend its adherents. There are fine people with religious backgrounds who remain loyal to their upbringing for various reasons. If these people see limitations they overlook them for want of thinking there is anything better. Some have loyalties borne of close friendships or family ties. But there is danger of having faith in a faith or faith in others rather than in the living God and His word. Pressure is sometimes exerted upon those who show signs of wavering. Then leaders can be selected because of their enthusiasm for what they had been taught and submitted to. Personal convictions can be suppressed through a sense of false humility. The more enlightened might have a sense of helplessness. Over centuries many would have been rendered intellectually impotent, fearing threats of death for heresy. One must feel thankful for one of the blessings of secular democratic governments which put an end to the temporal powers of religious despots. It was God himself, in his infinite wisdom, who put paid to theocracy by espousing King Nebuchadnezzar.

Some criticism is unavoidable. Jesus did well to describe self-important leaders as being like children playing in the market place and crying, "we have piped and you have not danced!" This should not be taken to imply that confessions of faith and catechisms lack substantial elements of truth. The generics of most systems can, however, be likened to the proverbial sound tree that is invaded by birds in its branches. Men have found it all too convenient to add to God's word on the grounds of tradition or to take away from it by a process of criticism. Self-opinionated views can be bolstered by taking scripture out of context. Constructively, a solution is seen as emulating people in Ezra's day. They repented of backsliding and syncretism after looking afresh at the law and discovering just how much they had fallen short. Reference will be made in chapter 4 and in the Conclusions and Summary to failed attempts in the past to repeat this.

Secondly, there is an obligation to show love. Jesus said to his followers that even enemies were to be shown love and that good should be done to those who despitefully used them. His ministry was both conciliatory and kind. He did not make a practice of attacking the beliefs of others except when he encountered religious hypocrites, and he was scathing about them. He found it necessary on one occasion to rebuke his followers for wanting to bring down fire on the Samaritans, saying they did not realise what spirit they were of. This was not the character of the finger of God. On the other hand, he made it clear that whoever persisted in following wicked ways could not expect to inherit the kingdom. People were free agents and God would not have it otherwise. His love for mankind is unlimited and it is spontaneous love that he looks for in return. He also expected his followers to show humility. The Apostle Paul found it necessary to remind believers at Corinth that some of them had previously been wayward. He wrote, "And such were some of you." Their transformation had been all of grace. They had nothing to crow about. Despite the humility of Jesus, religious leaders of his day took offence at his truthfulness. It is just as likely there will be adverse reactions to some passages in this book that are challenging but not intentionally confrontational.

Thirdly, there is a duty of care. One must not go beyond what is written. This is an easy trap to fall into. There is a general reluctance to admit any possibility of being wrong on any point. Christians who are frank admit, gospel apart, that some things are seen as through a glass darkly, although the gospel is clear enough, and difficulties can be avoided by keeping to impeccable early sources. It is taken that the writings of Mark and Luke were largely based on the witness of Peter and Paul respectively. Otherwise writings of post first generation believers are not relied upon. It is not suggested for a moment that implicit faith in Christ should waver or that there should be any doubting, but it is much better to face up to apparent problems than to ignore these, to remain silent, or to provide unconvincing solutions. Peter said that Paul wrote some things that were difficult to understand. It could also be possible that God has chosen to leave some matters shrouded in mystery. After all there were at least two mysteries hidden from the world until the first coming of Christ. The revealed word is inspired but not exhaustive. Many things were said and done that are not recorded. Thankfully, however, God has given to us all things that pertain to life and godliness as will be shown in chapter 5.

Then other areas are addressed. These are powers that oppose the kingdom such as the occult, hindrances such as false wisdom, passivity, and ambivalence. Satan's counterfeit manifestations are evidence of a supernatural realm. These things are covered in chapters 11, 12, & 13. These include authenticated accounts of frightening materialisations and the counterfeiting of such appearances. There is an example of youthful suicide due to involvement in the occult. There are absorbing accounts of the outworking of the finger of God in wartime; during Apartheid; in the deepest jungle; in a powerful tolerant, predominantly Muslim country; and in the course of meetings with a high profile African President.

The complex and controversial subject of Christ's coming kingdom is the central theme of Part Two. Due caution is exercised over matters that have caused even the best of scholars to disagree. Nevertheless, it is thought that interesting new light is shed upon the shape of future history and beyond. This demonstrates reasons why God, by his finger, will achieve a more perfect outcome of his plans for his creation than is allowed in most former studies although there have been one or two exceptions. These theologies of future things, or eschatology, are discussed in chapter 17, and both merit and anomalies are found in virtually all of these. One problem is that some seem to be designed mainly from a desire either to prove others' works wrong or to produce something novel. In some cases weakness is due to failure to encompass the whole historic evidence. For example, claims that the Babylon of Revelation was the ancient Roman Empire or later the Roman Catholic Church. Neither of these are sustainable because the first came to an end without the bowls of wrath being poured out or the two witnesses appearing. Attempts to identify individuals as the witnesses failed to recognise that these are to be killed and brought back to life with a resultant terrifying of the nations. This did not happen. As to the second, this is a religious institution and not a city that is the hub of world trade – which is what the future Babylon will be. Also some future harlot church existing at the end of time will be destroyed by the ruler of the world – the anti-Christ. The pedigree of this church is not known.

The author's approach is much the same as in Part One. The words of Jesus, particularly those spoken as part of his Olivet Discourse, are depended upon heavily. The revelations of the risen Lord to John are followed closely. A chronology of events is attempted rather ambitiously, and the role of the fall of Babylon disclosed. Faith in a restored creation in which God will win a total victory over darkness and see all his purposes come to fruition is expressed in chapter 21. A reason is given for the re-emergence of Satan from the abyss in the course of this final outcome. The rapture of believers and, as previously stated, Christ's second coming, figure prominently. In chapter 22, the culmination of all things is shown to involve a new heaven and a renewed earth, and an integration of perfected saints and angelic hosts among the converted nations. Key points that will be raised about the kingdom are: it is glorious and unassailable; it is about righteousness; the terms "heaven" and "God" are synonymous when used in relation to it; and the visible church is not the kingdom although many of its members have entered it. It is vibrant and radiant. Its earthly members have power but are no more exempt from conflict with the sinful nature than was the case in the early church. Historically, as more and more people set up their own denominations the voices of teachers became discordant and contradictory. Subsequent attempts at unity were frustrated when dominant factions were uncompromising.

It is argued that on the basis of logic alone there is greater likelihood that the finger of God created the universe than that it came into being of its own volition out of nothing. If the first hypothesis is acknowledged to be a possibility there is, surely, an implied responsibility to investigate it. As to the second hypothesis one should not be prepared to accept it purely on the say so of its advocates. Some secularist theories about life do not stand up to investigation but there is a great deal that is beneficial in the contributions of science to this planet's welfare. Science has been slow on the other hand to connect moral decline, and climatic pollution to long predicted portents concerning the destruction of this world as we know it, and threats posed by anarchical uses of nuclear fusion.

There is a spiritual reality which is a compelling alternative to escapism. Realisation of this should open eyes to a present kingdom at work in this world and that entry to it is available to all by means of repentance and faith in God. It is hoped that people might see that it is worth their while to learn about it and to seek after it. Many people are currently discontented with lack of permanence and the emptiness of secular philosophies. They are bored with the humdrum of everyday life. They seek to escape from it. They are attracted to imaginative unreality. Young children love fairy stories and delight in the innocent nonsense of Father Christmas. Nowadays the realm of imaginary unreality is booming. There are wide fields of science fiction, fictitious animal characters, and there are the burgeoning adventures of Harry Potter. There is a yet more subtle dimension in which it is harder for the gullible to detect whether events relate to actual or invented situations. *The Da Vinci Code* is an example of a totally fictional story that many chose to take as being true. This has proved to be a lucrative field because people like conspiracy theories and horoscopes. This material can detract from the person of Jesus Christ. Indeed some of it is designed for this express purpose. It has no basis in reality and people ought to question its bogus sources. An example is the blasphemy that Jesus was married to Mary Magdalene. In fact readers are asked to put themselves in the role of jurists both when reading this book and every other publication dealing with the subject of the supernatural regardless of whether it is pro or anti.

This preface posts the author's conviction that God makes overtures to people by his spirit and that in turn they can respond or not according to their predilection. It might be through his written word; by preaching, or encounters with other people. There are many instances, along these lines, described in this book.

Those who have found faith in the finger of God and seek to work in the present kingdom can be used in everyday occurrences. These might amount to appointments with destiny for those who are witnessed to. As an example, on a cold winter morning at 6.30 a.m., a man, having stood up in an accident and emergency ward all night, went out and hailed a taxi. Unusually the conversation turned to the taxi driver's boyhood when he attended Sunday School. He queried whether it was possible for all that he was told to have been made up. The answer was that much was well attested. The Apostle Peter wrote that he was not following a cleverly written up story when he told of the power and coming of Jesus Christ. He wrote, "We (Peter, James and John) actually saw him on that mountain." They were eye-witnesses to his majesty. He referred to an earlier experience when Jesus was transfigured before them and his face shone as the sun and his raiment was white as light. As genuine as this encounter was the doubtful would not be convinced of its reality because they had not seen it. That is why Peter stated that there was a more certain word of prophecy that people should heed as a bright light in a dark place, until the day dawns and the day star rises in their hearts. On arriving at the destination the driver was given a Gideon New Testament. Was this an appointment arranged by the finger of God?

This book may in itself be an appointment for some readers. After all, the prophet Isaiah wrote: "I was found by those who did not seek me; I revealed myself to those who did not ask for me." This verse was taken up by the Apostle Paul. Jesus said, "seek you first the kingdom of God and its righteousness." This statement becomes more immediately relevant if it is realised that the kingdom is seen as a present reality and not something to be aspired to in another life. From an individual aspect it draws attention to an existing narrow gate that must be entered. From a general aspect it speaks of actual dynamism. As a hope it points to the inexorable fruition of God's ultimate purposes. One day all God's people will be one. One day they will serve and work enthusiastically in his kingdom. It is only known for certain that this hope will be fulfilled in Christ's coming kingdom.

Note: As this book was in the course of editing, information came to light from a surprising source that underlines a major contention and concerns the Lord's Prayer. This has been included in the Conclusions. It provides a compelling finale.

PART
ONE

God's Present Kingdom

1 | The Nature of the Kingdom

"The Kingdom of Heaven is not meat and drink but righteousness and peace and joy in the Holy Spirit." Romans 14:17

There is only one kingdom of God both as to time and to extent. It is a habitation of those who have entered it, and also prospectively for those who seek it. This kingdom comprises all spheres in which the righteous rule of God prevails. Its present territorial occupation extends over eternal and temporal habitations. This oneness of the kingdom can be comprehended more clearly if the terms "heaven" and "God" are understood to be synonymous when used in conjunction with "kingdom." On the temporal plain its characteristics do not resemble finite objects (meat and drink) but consist of unseen and eternal principles (peace, righteousness and joy in the Holy Spirit). It appears, as we shall come to, that the eternal or heavenly sphere only differs in that it possesses, additionally, visible dimensions of unimaginable lay-out. In order to comprehend a model of a spiritually perfect kingdom one might begin with a process of eliminating everything that is evil and envisage all the attributes of goodness that might then reign alone.

Thomas More wrote a little classic entitled *Utopia* in which he described an imaginary perfect state. The frontispiece of the 1923 Modern English Translation read: "The working part of the population of England carries a mass of non-workers on its back all the while . . . We cannot exactly afford so many idle hands; nor can afford the number of empty minds England has today." There is (in 2008) still this marked social idiosyncrasy except that the idle hands are different. The idle socialites and aristocracy have had their affluence dented by death duties and rising wage costs and those who are now the idle ones come from other backgrounds. These are lottery winners, benefit scroungers, malingerers, the perpetrators of scams, casino owners, and a new executive class that control companies and public services. The lesson here is that the composition of darkness shifts whereas God's righteousness is unchanging. More's "Utopia" was so popular that this word entered the English language as "an ideal perfect place or state of things." Hence "Utopian" and "utopianism". In Utopia most people had faith in an unknown Divine power, eternal, incomprehensible, inexplicable, far beyond the reach of human intellect diffused throughout the universe in power and potency. This is uncommonly like the Holy Spirit of which Jesus said, "The wind bloweth where it listeth and you do not know from whence it cometh or whither it goeth – so is everyone that is born of the Spirit." He added that except a man be born again of water and the Spirit he could not see the Kingdom of God.

Those who enter this kingdom will display its attributes, or fruit, arising from the operation of the Holy Spirit. Love, joy, peace, patience, kindness, goodness, faithfulness, gentleness, and self-control. This is as opposed to the works of the sinful nature that are epitomised in the world. These are: sexual immorality, impurity, and debauchery, idolatry, witchcraft, hatred, discord, jealousy, fits of rage, selfish ambition, dissensions, factions and envy, drunkenness, orgies and the like. These lists give an understanding of the nature of God's temporal territory and what is opposing it. The sinful characteristics give rise in people to swindling, killing, stealing, mugging, vandalising, looting, engaging in acts of terror, lies, anarchy, and abuse of others and ones own body. At corporate level the same characteristics engender wars, genocides, oppression, exploitation, pollution and avarice. Now the work of the Spirit in the world is to combat evil and within believers hearts it is to conflict with sinfulness.

The sword of the Spirit (the Word of God) is a weapon of spiritual warfare. In non-believers the Spirit causes conviction of guilt in regard to sin, and righteousness and judgment. The eternal territory is exempt from sinfulness. It is impressively huge, being populated by hosts of angelic beings serving the triune God together with a great multitude that nobody can number out of every tribe nation, tribe, people and language. The latter are an assembly of departed souls whose trust was in their maker and his Christ. After a lifetime's conflict with their own personal defects and opposing the world's darkness is over, they enter a sphere in which God's righteousness reigns and they experience delights that are beyond human comprehension. Eye has not seen nor ear heard neither has it entered into the heart of man what things God has prepared for those who love him, but he reveals these by his spirit.

In a supernatural experience the Apostle Paul was shown something of the heavenly realm that proved to consist of things that were unutterable. This makes sense as with God there are surely dimensions, colours, and parameters of comprehension that are beyond our limited understanding. The world's cleverest brains still have only a limited understanding of infinite creation even after access to accumulated knowledge. It also accounts for the use of apocalyptic language by the prophets like Ezekiel and the Apostle John in order to describe heaven. This is no reason for criticising them or for doubting the authenticity of their descriptive passages portraying visions or to question apparent limitations. There is clearly even a heavenly vocabulary described as "the tongues of angels," which would provide a common vehicle of communication rather like the divers tongues of Pentecost when people understood perfectly what was being said, each in their own language. This was a momentary overturning of the confusion of tongues imposed on earth at the Tower of Babel, and a foretaste of the eternal city. These were not "unknown tongues."

As will be set out in chapter 2, all justified human beings who lived before the death and resurrection of Jesus had to await in Abraham's Bosom until these momentous events opened up the heavenly territory to them. They would be joined by all deceased believers until the end of time. After this there will be a destiny for all, to be discussed in Part Two. However, at this juncture the nature of the present earthly Kingdom is of more immediate importance and it will also be useful to include a brief account of the pre-kingdom epoch and the non-kingdom part of the current temporal scene. As individuals we live on this planet just for a lifespan. We become aware of the world around us and of peoples and cultures. We are influenced by our families and our ethnicity. We also have the benefits of accumulated wisdom through education. We are confronted by a complexity of governments, utilities, companies, organisations and religions as well as secular philosophies. Despite the tenuous nature of life attended by famines, pandemics, earthquakes, tsunamis, wars and crime, there are opportunities to acquire wealth and be acknowledged by others. This all describes the "world" as meaning the society in which we live as opposed to a geographical location. The former sense applies to the saying of Jesus that his kingdom was not of this world, meaning not of it although in it. The nature of this present earthly kingdom is of overriding importance because, working in the world, it is at the sharp end. It is here that good combats evil and light confronts darkness. It is here where all kinds of people gain admittance to eternal life and enter the kingdom. This also means that in the here and now a place is reserved for them in heaven. It is a crucial testing ground; an arena of great spiritual opportunity. Publicans and harlots press into the kingdom but religious hypocrites do not and sometimes prevent others from doing so. A hymn writer called the temporal kingdom "Immanuel's ground": a terrain that is passed through to fairer worlds on high. This is a dynamic territory. As Mary said of her son, "He has filled the hungry with good things but has sent the rich empty away." It is light that has come into the world but it is not always preferred because men love darkness rather than light and do not want to come to the light where their deeds will be reproved. One's own righteousness falls very far short of God's holiness.

The problem with organised Jewry was that it went about to establish its own righteousness by the law being ignorant of God's righteousness imputed to Abraham as he believed in God. It is salutary that not all professed Christians have the faith of Abraham; a faith that is a necessary factor in gaining admission to the kingdom. The parable of the dragnet makes this abundantly clear. Jesus said the kingdom of God was like a net that was let down into a lake and caught all kinds of fish. It contained both good and bad fish. The bad were thrown away. This was an allusion as to how the righteous would eventually be separated from the wicked. In the meantime both of these types will remain together in the visible church. This demonstrated that when Jesus referred to "My church", he only had the good fish in mind. That is those who had been met spiritually by the finger of God. They had not looked about for the Kingdom (not said, Lo it is here or Lo, it is there by observation) but had found it in their hearts.

Status is unlikely to figure in eternity although there will be privileges and rewards. The parable of the labourers in the vineyard gives an insight into the divine mind. They received every man a penny and those who had borne the heats of the day grumbled. However as the penny in this case was meant to represent the gift of eternal life a different complexion can be put on the story. As an example, in the Sultanate of Brunei people are unconcerned that it is not a democracy. On account of its natural resources and sagacious foreign investments every citizen is given a house, a car and permanent employment and pays no income tax. This is a utopian state but it pales into insignificance when compared with the unsearchable riches of Christ in glory! What a present boost to hope it is to realise the nature of the kingdom and that its consummation is God's uttermost and divine purpose! It will be a masterstroke and the denouement of his strategy. Present blessings that are enjoyed such as new life, fruitfulness, empowerment, and fellowship in the Spirit, together culminate in the contemporary ingathering of souls into the eternal kingdom wherein dwelleth righteousness.

The following chapters address the subjects of conflict on the temporal territory and joyful and sorrowful experiences of earthly pilgrimages. Jesus said that in the world God would set good seed and Satan would set weeds and these would grow together. Since it was often hard to pull up weeds without disturbing the wheat or to spot the difference it would be necessary to let both grow together until the final harvest. Apparently members of the amorphous population can be indwelt at some point either by God's spirit or by the Devil's planting. Hence the ongoing battlefield of light against darkness. Citizens in the course of this will undergo pruning, testing, chastisement and persecution. Strangers will be convicted of sin, judgment and righteousness. God does not will that anyone should come into condemnation and join those who are condemned already: namely Satan and his wicked angels. Citizens can be drawn from people of any denominational background or those with none. There will be no labels in God's presence.

Today it is necessary, socially, to declare one's religious background to be Catholic, Orthodox, Anglican, Methodist, Baptist, Pentecostal, and so on. Nationally this is called for by hospitals, the armed forces, and censuses. Chaplains need to know who belong to their flock. There are also different confessions and traditions that straddle denominations. It is not general for persons to enter "Christian" on documents. If such a single nomenclature could be adopted universally it would give less confusing impressions to those of other faiths or those who have none at all. It seems impractical to take such a step towards unity. However if a wider acceptance of promoting citizenship of God's present kingdom on earth was encouraged there would be less confusion on the part of strangers and a greater sense of unity across denominational boundaries. This proposition will be discussed again in the Conclusion. Those nations who admit immigrants see a need to teach the indigenous language to foreigners and to explain customs and culture. So let citizens teach the language and culture of the kingdom to new entrants so that they grow up into Christ. There is almost always an instant affiliation between real Christians even when meeting for the first time. It is an added blessing when they meet consciously as fellow citizens of the kingdom of God.

2 | History of the Kingdom

"Your kingdom come, your will be done on earth as it is in heaven."
Matthew 6:10

The origins of the Kingdom of God cannot be fathomed because it is an everlasting dominion. Undeniably, God was always master of eternity, and there is evidence of certain pre-creation events. A triune God dwelt in a glorious isolation and at some stage created a host of angelic beings. Such is the divine attribute of royal kindness it follows that all subjects are always given options to pledge loyalty or not. God is not only love but yearns to receive love spontaneously from others. What would be the worth of exercising compulsion? It would seem that this generosity was the root cause of Satan's attempt to set himself up as superior to God and that in his revolt he led a number of angels with him. It is hardly the business of humankind to pry too far into the timing and outcomes of pre-creation heavenly crises or to attempt to plumb the depth of Satan's cunning. For instance, it would be pure conjecture to suggest he was jealous about the introduction of an earthly creation and of Adam's delegated command of it. Or was he simply aiming to spoil God's great counter-attack of substituting a human race to love and be loved by it?

The divine finger created the universe and furbished this planet. God was pleased. This was no skirmish or pilot study. He walked in the garden with Adam and Eve but this was not the kingdom of God. There is no certainty about the status of wicked angels before Christ's advent. It is known that after the Fall of humankind Satan still had access to God and could bring accusations. There was never any question, however, that evil ever had controlling authority. Satan was able to bring a temporary setback to the plan of creation by deceiving the first humans and bringing about their disobedience. As a consequence what had started out as a perfect world became degraded and marred. During the preceding period of innocence there was no killing as far as any of the species was concerned. There was no carnivore, no disease, and no death. After the fall a great rescue plan had to be activated so that something infinitely better than a mortal paradise would emerge in due course. In a nutshell, the human race would have an option of being transformed into God's own likeness as by the power of The Holy Spirit. More will be found about this in Chapter 5.

Historically mankind failed because of disobedience in the garden, and then failed to keep the law given to Moses they had vowed to uphold. God agreed reluctantly to accede to their pleas for a visible king even though at the time Israel was the world's sole theocracy during the amphictyony. King Saul was a disaster and, therefore, a continuing theocracy became precarious as successive kings either did evil or good in the sight of the Lord. In his infinite wisdom God later gave secular power to Nebuchadnezzar, but never again was he to resume direct worldly governmental responsibility, his kingdom no longer being of this world. So much for the pre-kingdom epoch, except that well before the exile in Babylon God had initiated a new plan for a greater destiny. Humankind was to have opportunities of inheriting his kingdom. This plan began with his approach to a man called Abram who was re-named Abraham. Four hundred years before Moses was given the law God told this man to leave his own country in the Ur of the Chaldeas and go to an unknown destination. He believed God's promise that he would become the father of many nations: a promise that over time was fulfilled by the descendants of his sons Isaac and Ishmael. Much more importantly, Abraham was looking forward trustfully to a city with foundations whose architect was God. That city is what is here being referred to as the eternal territory. Abraham believed God's promise and it was counted to him as righteousness.

This Abrahamic faith was identical to that exercised by every person who has or will in the future become fellow citizens of the kingdom. From the time of Abraham to the emergence of John the Baptist from the wilderness the faithful were looking forward to the promise to be fulfilled in a Messiah who would open up the kingdom. It had been foretold by the prophets. There was something of a veil over the exact manner and detail of how expectations would be realised. Some thought after the captivity of the Jews in Babylon there would be a restoration of the earthly Israel. Jesus was asked, "will you restore the kingdom to us?" Some sects still believe this to be a literal future event in present world history. These are mistaken because any future restoration will be beyond history and will encompass both Jews and Gentiles with the faith of Abraham. Jesus said that some of the natural descendants of Abraham did not have God as their father.

The true kingdom of God that John the Baptist heralded was initially imminent during the earthly ministry of Jesus. In order for the kingdom to come Jesus was empowered by the finger of God to fight great spiritual battles against Satan and all his forces of darkness. Jesus mastered Satan in the wilderness, overcoming temptations by the use of the Word. He bound Satan and plundered his house; he took away his armour; the deliverance of humankind was wrought at the cross; the kingdom was opened and Jesus, through death, destroyed him that had the power of death, that is the devil. The simultaneous rending of the curtain in the temple signified that in Christ there was no longer Jew or Gentile or separation of one nationality from another. All could henceforth become one in Him. On that day many of the just dead who were buried in Jerusalem were seen walking about in the city. This was, undoubtedly, a further sign that the kingdom had been opened to humankind. Satan could no longer bring accusations against the tribe of Adam after Jesus saw him fall as lightning from heaven. From thenceforth demons could only operate on temporal territory. They were already judged and knew that they had no future. They pleaded with Jesus on one occasion not to be disturbed before their time. So he cast these out of a man and they entered into a herd of swine who stampeded over a cliff and perished.

Living people were entering the temporal spiritual kingdom violently for the first time. Initially these people spanned two eras. For instance, Simeon and Anna had been given an assurance by faith that they would meet the Messiah before they died. They encountered the infant Jesus in the Temple and thus saw the Son of God in the flesh. They then had a relatively short wait for transport to the eternal kingdom at death. So the kingdom of God embraced all those with the faith of Abraham at the precise time of Christ's victory regardless of whether they were still living or waiting in Abraham's bosom. Those who had already departed were escorted by angels into the eternal heavenly kingdom. They were the first human occupants. Those who still remained on earth had a place reserved for them in eternal territory and were already citizens actively serving in the temporal kingdom.

Many theologians did not believe that the kingdom of God came on earth. It was either still imminent or future. They gave explanations to justify such views instead of believing Jesus. One even claimed that Jesus was mistaken. He ran a leper colony but this cannot have profited him without love. Another cited twenty-nine instances in the New Testament in which the kingdom was stated to be future, thirty-one without notice of time, and sixteen referring to it as present. His odd conclusion was that since there were more references to the future then the kingdom had not come! Happily there have been many outstanding men who have taught that the kingdom was both present in the world invisibly and spiritually, and that one day it would become visible at the second coming of Jesus. What seems to have daunted the rest is that, like the Jewish nation of old they could only contemplate a visible kingdom in open glory. And, of course, they overlooked that God had finished with theocracies. This confusion has not helped the visible churches to show a consistent front. Rather it has been a factor causing divisions.

Whenever the visible church had overriding power over judicial and constitutional affairs it failed woefully to manifest the spirit of Christ. It dominated, persecuted, and practiced violence. Whenever it was persecuted its sufferings and martyrdoms were glorious: it epitomised the spirit of the true kingdom. This chapter is concerned with the history of the invisible true church. It is neither pleasant nor edifying to dwell upon the failure of the visible church. Its sad history will only be recalled here to bring these two churches into juxtaposition with each other and likewise in chapters 3 and 4 as a warning about how elements of the visible church have hidden, and will continue to hide, the kingdom from others. It is essential that this negative influence should not be ignored altogether.

God's intention to bring the kingdom down to earth was something that he hid from the beginning of the world. It remained a mystery until the fullness of time had come. Events on the day of Pentecost were another visible confirmation of its arrival. It was only after Pentecost that believers began to meet together in any numbers for prayer, fellowship, and the breaking of bread. This they did from house to house to start with. As the response to the gospel message grew the notion of an ecclesia emerged. The natures of the seven churches mentioned in Revelation confirms this. In the beginning believers were under threat from the outside. The history of the catacombs confirms this. Later when outside persecution ceased, the threat to the truth came from those within. The birds in the parable of the mustard seed, already mentioned, also represent wrong doctrines. As visible churches grew they were penetrated increasingly by false teachers. There is unlikely to be a totally pure visible church. It is the hidden church of true believers whom Christ loves and gave himself for who are sanctified progressively by grace and the abiding word. One day this church will be presented to himself without wrinkle, spot, blemish or any such thing. Only "the birds" claim that a visible church can be pure. They do this as a means of self-justification. The ancient church of Laodecia said, "I am rich; I have acquired wealth and do not need a thing. "But Jesus said of it, "You are not rich, you are wretched, pitiful, poor and blind." That church needed to partake of the divine nature.

Some of the epistles were written partly with a view to present truth in such a positive manner that the growth of heresy would falter. But as the birds of the air multiplied so the use of the term "heretic" became seriously misused. Those who dared to fight for the maintenance, or in some cases the re-emergence, of biblical truths were themselves branded heretics. This was when all the time the real heretics were the ecclesiastical suppressers. A great tension developed between the simplicity of the gospel and the complexity of proud theological inventions. This was akin to the tensions that developed between Jesus and the religious authorities of his day. Furthermore, it was a matter of the light of the kingdom being opposed by the powers of darkness. It can be truly said that if the light that is in people is the darkness of religious tyranny: "how great is that darkness!" This darkness was so beguiling that it attracted those who loved the pomp and splendour of ornate symbols, of pompous ceremonies, and gave undue reverence to people and elevated them improperly instead of giving glory to God alone. There was something about all this that drew great admiration from carnal minds. Once the spirit of God left a church the remaining religious environment was considered to be the next best thing. It was also something that was defended by cruel means for fear that gross falseness might be unmasked. For instance, there was such fear of teachings about the second coming of Christ that the bible began to be challenged as to it's accuracy on aspects of prophecy like this. It might well be asked where all the self-proclaimed false leaders of professing churches will stand in the face of the coming King? In the meantime one argument is that the Book of Revelation is not in its correct place, and that a different chronology would diminish its importance. Another is that it should not be in the canon of scripture at all. Whatever the arguments, it is a fact that recognition of the second coming was lost for many centuries by the broad church. Apparently, a Spanish priest in South America was the first to bring this back to attention.

An interpretation of a dream by Daniel showed a succession of kingdoms until finally at the very end God's kingdom superseded all of them. The coming of this final kingdom will usher in Christ's visible reign. The purpose of the present invisible earthly kingdom is to bring light into the world and many sons and daughters into its citizenship, and to fight against darkness. Therefore, as well as showing that historically there was a time difference between the imminent kingdom and its inception on earthly territory, a further time difference needs to be noted between the latter's inception and a visible kingdom coming beyond history. As a reminder, the powers of the kingdom were present during the earthly ministry of Jesus from the moment the Holy Spirit came upon him at Jordan. Although the kingdom was at hand no person could enter it until the victory of the cross when Satan was defeated. What came was a kingdom that cannot be shaken. It is only a kingdom that is entirely spiritual, and not tangible, that can sustain its existence against the shaking of the heavens and the earth.

So, although we do not know anything about the pre-creational, eternal, infinite, and everlasting chronology of the kingdom of God we have been able to pick up its history at some point and from this time be fairly positive about events that have involved mortal beings. We have also seen the magnanimous and merciful nature of God in making the entry of our forebears retrospective. As to the future, we can confidently expect the exercise of the same magnanimity and mercy as God does not change. The future is not obscure in the same way as the distant past because we have the benefit of the sayings of Jesus and the prophets. There are a group of psalms (96 to 100) that speak of the future kingdom. The prophet Isaiah devoted chapter upon chapter to a future universal kingdom. The Olivet Discourse of Jesus gave a succinct summary, and then in his risen power Jesus revealed a great deal to the Apostle John as recorded in The Book of Revelation. It is sufficient to note that the final kingdom will be a place wherein complete righteousness dwells. The detail will be found in Part Two.

3 | The Hidden Kingdom

In both this chapter and the following chapter ways are explained in which the true nature of the kingdom can be hidden from mankind. As already stated, it has been found convenient to separate people into the categories of "stranger" (one outside of the kingdom) and "citizen" (one already inside) although, as will be realised, there are some factors that apply to both. This chapter applies predominantly to the first mentioned.

Hidden from Strangers

(1) The god of this world. A common trap is avoided. That is by not blaming religious divisions and the historic notoriety of some denominations primarily for antagonising and alienating ordinary people. In fact, the greatest hindrance is the god of this world and not religion. For example, the Babylon of Revelation is claimed to refer to Roman Catholicism. That this is untrue is shown in chapter 20. In fact this Babylon is a city that is the hub of the trading kingdoms of this world. That its characteristics are those of unrighteousness is evident from the current greed, exploitation, and dishonestly being exposed day by day in the media. Those who serve these kingdoms are blinded by the god (note small 'g') of this world. Corporately, the aim of the captains of these people is to acquire material wealth, power, prestige in the eyes of others by any means. They can be readily identified by a desire to live in luxury and to enjoy personal pleasures regardless of the needs of others. Deceitful financial practices, and secret dealings and bribery are rife. Fraud is taking place on a large scale. Those involved in these things have no time for the kingdom and its righteousness or for those who have.

On a personal and less culpable level worldly activities can distract people from seeking the kingdom. Not that worldly activities are wrong in themselves. It is possible to pursue these, however, to the exclusion of consideration of spiritual matters: of asking why we are here and whether there is an eternity to be faced. Recreation and exercise are healthy. These keep the young out of less appropriate activities. Not so much can be said of some aspects of spectator sports especially those involving multi-million pound merchandising of players, huge sponsorship deals and the use of anabolic steroids. Ordinary social activities are part of civilised life but if these are pursued to excess the kingdom of God can be relegated to a side issue or a non-issue. Then there is peer pressure. People can be reticent if facing the probability of mockery in the workplace or when associating with acquaintances. The draw of worldly friendships especially in groups is powerful and sometimes leaves very little room for independent thought or behaviour. Another challenge to genuine personal opinion is the media. This becomes more and more partisan and ranges from newspapers, television, and other audio vision and audio channels. Many come to believe that views read or heard are their own such is the power of journalistic and commentator persuasion.

(2) Those who are lost. The bible states that the kingdom of God is hidden from those who are lost. The word "lost" should not be construed as meaning this has to be a final state. Jesus came to seek and to save those who are lost and it is open for anyone to seek for themselves. There are many reasons why the kingdom of God is hidden. It is not easy to enter. Its gate is narrow and it is as hard for a rich man to find it as it is for a camel to go through the eye of a needle (that is to go through a rocky passage when side-loaded with baggage). However, with God all things are possible. A narrow gate and the eye of a needle are images that are underlined by the account of the Prodigal Son. Wealth came to him very easily and this resulted in dissipation. When the money ran out the profligate came to his senses. This is not an unfamiliar scene today. Drug addition is the result of wealth, perhaps inherited or stolen. It is more sinister in that Satan is often instrumental in causing suicide or death through ruined health, as will be shown in chapter 11. A further problem the lost have is in finding the correct information when they start to seek.

(3) Proliferation of information and disinformation. A truth can be hidden amongst a welter of other information or disguised by misinformation. A hidden thing is something that exists. If it is a tangible object searchers will try to find it. If it is a truth searches will need to unravel it. A secular example is a problem that President Franklin. D. Roosevelt was confronted with. He was criticised for failing

to identify an item of intelligence about Japan's intention to attack Pearl Harbour in 1941. However, it was all too easy to be wise after the event when at the time the item was but one of thousands of pieces of intelligence. Some were irrelevant and some were bizarre. For example, in Hawaii a dog on Oahu beach was reported to be "barking in Morse code" to Japanese submarines off shore! Most items were of a more rational kind and so obscured the vital one. Some junior signaller may have been convinced of its validity and urgency but then he would not have read all the other messages. One has only to reflect that there are two thousand nine hundred different religious sects in America alone to be able to apply the Roosevelt incident to looking for the kingdom.

On the matter of disinformation a double of Field Marshall Montgomery was sent overseas as a masquerade in order to deceive the Germans into thinking that an invasion of France could not be imminent. So too in the spiritual realm strangers are faced with contradictory and false information. Then some scientists say there is no God and our universe was a self-starter. There are the claims of evolutionists that man is descended from monkeys. There is the teaching of Calvin who thought God was making a monkey of mankind by devising sin and judgment and making him powerless to do anything about it. Well attested events both by prophets and eye-witness (such as post-resurrection appearances of Jesus) are dismissed as "pie in the sky", whereas stories known to be fabricated are accepted as being true by gullible readers. Books such as *The DaVinci Code* and *The Bible Code* are recent examples.

(4) Complacency. The full meaning of some of the parables of the kingdom uttered by Jesus seemed to have been hidden from his hearers. He explained these afterwards to his immediate followers. It has been conjectured that, firstly, it was too dangerous for him to go public at that stage, and, secondly, that it was inappropriate to tell all the people everything. A more detailed study will show that people were dull of hearing and forgetful. Jesus only appeared obscure because he had already laid matters out plainly. His fuller explanations to his followers of a subsequent parable have been of benefit to many generations through the inspired page. If one studies closely all that Jesus said and did he or she will not only have a sure guide about the kingdom but also a clear understanding of other fundamental matters that hinder the complacent. It can be seen in John chapter 5 that earlier Jesus spoke with the utmost clarity. He said that eternal life came though his word, and a time was coming when the dead would hear his voice and come out of their graves to live forever or to be condemned. He had authority to be their judge. He stressed that he did nothing of himself; only what the Father showed him. By himself he could do nothing. How refreshing and different this was by comparison with those who in future ages perpetrated their own ideas. One has to be alert about this and stick to what Jesus said. If information is not in the bible even it is alleged to be Christian tradition it must be regarded as suspect. It can hide the truth of the kingdom. The thirteenth chapter of Matthew's gospel shows clearly that nothing was withheld overall. It was just that the Jews were careless. This is a lesson. Jesus quoted the prophet Isaiah, "You will be ever hearing but never understanding; you will be ever seeing but never perceiving. For this people's heart has become calloused; they hardly hear with their ears, and they have closed their eyes. Otherwise they might hear with their ears, understand with their hearts and turn, and I will heal them." One is in no better position than an atheist if subscription is to half truths and distorted truths that do not lead to the narrow gate. It is sad that people rely on what they think is an insurance policy to cover a ticket to heaven and are unaware that it is bogus and keeping them from the true means of obtaining an incorruptible inheritance reserved for them that does not fade away.

It is a mistake to accept every view even of an acknowledged stalwart of the faith on the grounds that if he believed something, "who I am I that I should disagree?" That man of extraordinary faith, George Muller, believed erroneously that it would have been impossible for him to resist God's grace. Cardinal Newman did not believe in Transubstantiation until he went over to Rome and acquiesced out of a sense of false humility.

(5) Poor witness. Another strong reason why the true kingdom is concealed from strangers is the disastrous witness of some Christian leaders. These are either entirely false or backslidden. Because all religion can be seen wrongly as synonymous with the true kingdom, people, understandably, become scandalised by such things as sectarian violence and autocratic dictatorship perpetrated in the name of righteousness. Then there are preachers who commit adultery, or are immoral in other ways. If an ordinary Christian falls into bad company his or her fall is seized upon with glee, and the gospel thereby decried, by those who have no qualms about their own behaviour. It is not against the teaching of the bible to drink wine. In Wesley's day adults drank ale because the drinking water was polluted. The Victorian total abstention lobby attacked the social evil of cheap drink and subsequent rampant drunkenness that ensued. This became confused with biblical teaching. Temperate consumption of alcohol is not sinful. But drunkenness is. Most Christians abstain so as not to stumble the weak minded or those who are addicted. It is a matter of good witness.

(6) False security. Another dangerous way in which the kingdom can become hidden is by doctrines that lull beings into a false sense of

security. The true shepherd enters by the door of the sheepfold. Whoever comes in by any other way is a wolf in sheep's clothing. Probably a sheep shearer. No ritual in itself can make one a child of God and an inheritor of the kingdom. Doctrines of purgatory, prayers for the dead, and indulgences clash with the reality of a present kingdom. It is not proposed to discuss motives. It is more vital to point out that to give undue weight to the powers of water baptism and other sacraments by making these a means of grace in themselves will cause people to rely almost entirely upon symbolism without regarding the essential role of The Holy Spirit in the regeneration of a soul. Weight of scripture makes traditional views like this seem light. As has been touched upon there are also elements of Protestantism that are misleading. Calvinistic doctrines laid down at the Canons of Dort – a state convened assemblage – declared that God predetermined who would be in and who would be excluded from the kingdom; people had no responsibility for their own destiny; that Christ only died for some; that God's grace was irresistible, and that man is totally depraved. In that this is fatalistic it encourages either a false sense of security or a sense of hopelessness. Calvin regarded his five points as a "monstrous nightmare" and stated that God's decree was horrible. Why couldn't such an acknowledged scholar have conceded that there might be another explanation more in keeping with God's attribute of love? Jesus invited all to come to him. Why ask people to repent if they are unable to? God does not want anyone to perish, but that everyone should come to repentance.

Then Arminians teach that it is possible to be in the kingdom one moment and out of it at the next. Now this is the reverse of false security but the teaching was bred through a dislike of such and is worthy of sympathy but not belief. It was borne out of objections to "easy-believe-ism" – a presumption that is grievously false because not all who say Lord, Lord, will enter the kingdom. Jesus was so clear about this in his parable of the sower. When the seed was scattered some fell on the paths and the birds came and ate it up. Some fell on rocky places where the soil was shallow. The plants sprang up too quickly and the sun scorched them and they withered because they had no roots. Other seeds fell amongst thorns which grew up and choked the plants. Still other seed fell on good soil, flourished, and brought forth abundantly. This parable as so far related by Jesus was not understood. Fascinatingly, Jesus told his disciples that the secrets of the kingdom were only for them and not for the crowds. Clearly, there was something in this teaching that was inappropriate for all and sundry. It would only give way to busybodies and judgmental situations. Only the Lord knows those who are his. His explanation was that the seed represented the word of God and the parable was about how people reacted to it in different ways. The first three instances were about people who were unfruitful and not, therefore, genuinely regenerate. In the final case an abundance of fruitfulness was evidence of divine acceptance. Jesus said that by their fruits you shall know them. These were those who had entered the temporal kingdom and were bringing forth the fruits of righteousness that are in Christ Jesus. So to believe that by "going out the front" or putting one's hand up and making an empty profession one is saved is just as false a sense of security than believing in the effectiveness of purely ritualistic water baptism.

(7) Denominationalism. It is obvious to an outsider that since there are different views only one can be right although it does not follow that any one denomination will be right or wrong about every view. This leads to disenchantment about spiritual matters and to incredulity about claims of infallibility. So in that a denomination has got things right it is conforming to the metaphor of the mustard tree. In that has been infiltrated by tradition and extra-biblical teachings or has taken things away from the word it is symptomatic of the birds of the air. Only recently a new Pope, who is a respected academic, declared that his church was the only true denomination in Christianity. This statement has disturbed several other denominations. The Pope has failed to see the speck in his own eye (something we can all fail to do). There should not have been surprise because although the 1962 Council (Vatican II) made overtures to "separated brethren" and expressed regret for past confrontations, a careful reading of the Council's documents would reveal that all along Rome only sees one way to unity. That is to submit to the Primacy of the Pope and all his teachings. As the Roman church, over the centuries, added to the Nicene Creed up to seventy extra-scriptural teachings on the basis of tradition, and then lost several embarrassing ones by stealth, it can hardly claim in all rationality that it is more pure than other denominations. In any case infallibility is only a recently promulgated doctrine. It is paradoxical that the Roman Church has on the one hand been a glorious champion of the doctrine of the Virgin Birth by helping to preserve this key facet of faith and on the other has introduced extra-biblical doctrines about the Blessed Virgin to bolster its credibility. For instance, that Christ will one day present to himself a perfect church *through her.* To suggest this is to hide the true nature of the coming kingdom. Teaching about an immaculate conception of Mary that Thomas Acquinas taught against, and bringing in a modernist fashion of worshipping the sacred hearts of Jesus and Mary and Joseph, that the Gallican Church successfully delayed, contributes to unwelcome detraction from the uniqueness of Jesus whose glory must not be given

to another. Moreover it hides kingdom values from strangers.

The intention is not to grind an axe with Catholics, many of whom are splendid and humane people, but rather to show that claims of supremacy have to be refuted in order to reveal the nature of the true kingdom to those who are lost. At individual levels all kinds of people in the world are affable. The Pontificate is no better and no worse than a lot of the world's cabinets whether these are national or global, secular or religious. Whenever powerful agencies exert their wills over a powerless populace this is Babylonish: unlike Abraham Lincoln's vision to provide under God a government of the people, for the people, by the people.

(8) Unravelling prophecy. People become baffled by prophetic writings. These are difficult to unravel but a greater uncertainty can be blamed on contradictory interpretations. Farrar nominates men from the distant past as the two witnesses of Revelation but these did not come to life again and, therefore, could not possibly have filled the bill. Jesus himself was uncertain about the precise time of the consummation of the kingdom although he spoke of events that would immediately precede it, and how these would develop, and so did Peter and Paul. It could be that the time of the end will depend upon the completion of other events and that is why it is hidden. After all God stopped the sun in its tracks for three days in Joshua's time. All the other separate events are set out clearly enough and the bible has not been caught out. The prediction that the world would get hotter and be ended by extreme heat or fire seems well on course for fulfilment. In order to overcome confusion a feasible and comprehensive explanation, free of bias, and tested logically should be prepared. It should discuss the future portents of the tribulation; a second coming; rapture; a millennium, and the consummation of the present world as it is known. This is attempted, not without trepidation, in Part Two.

(9) Doctrines of Men. The teachings of Jesus and his Apostolic disciples are unequivocal on all points relevant to salvation and life. Therefore, any confusion of strangers must be due to either magnification of less important details by critics, or the introduction of extravagant doctrines by pseudo intellectuals. As to the former it takes the humility of a Paul to admit that here we see as through a glass darkly (but then – in the future – face to face). He also wrote that if people thought differently God would reveal it to them. Jesus instructed his followers not to obstruct a man who was preaching the kingdom independently. So reasonable ignorance and diversity is not troublesome. If there is a hungering and thirsting after righteousness then a person is not very far from the kingdom. However severe damage is done to the true message when people are deceived into

accepting error that detracts from the glory that is due to Jesus alone. The source of this deceit was clear to Paul who wrote, "...just as Eve was deceived by the serpent's cunning, your minds may somehow be led astray from your sincere and pure devotion to Christ." Doctrines that ascribe power and status to people who are in heaven that are the prerogative of God alone are suspect in origin because there is only one mediator between God and man, that is Christ Jesus. It is proper that humankind be acknowledged and honoured in line with biblical teaching. The Virgin Mary was willing in the day of God's power and was, therefore, described as the Handmaid of the Lord. Henceforth all nations would call her blessed. Extra-biblical doctrines were introduced stating that she was Queen of Heaven and a mediatrix. These doctrines caused one branch of the visible church to be guilty of over-adulation and another to underrate her entitlement. The point here is that conflicting doctrines get in the way of urgent kingdom business and leave Satan less opposed as he goes about like a roaring lion seeking whom he may devour. He should not be given the slightest foothold in his attempts to snatch away the seed of the word. Resist the Devil and he will flee from you.

(10) Questioning God's love. Acknowledgment of the greatness and goodness of God must always be paramount if the kingdom is to be presented to strangers with conviction. There are those who query his title of God of love because of the destruction of people in the deluge and at Sodom and Gomorrah. Firstly, it is wrong to question God. Secondly, there must have been perfectly good reasons for his actions. Conversely, God's decision not to intervene directly in this age is construed by some as due to remissness on his part when it is more likely that end times are near enough to make any action inconsequential. Prior to the deluge Noah preached for one hundred and twenty years through the Spirit of Christ that was in him. A genuine response to his message would have guaranteed eternal life but probably not earthly survival. The human race had become corrupted through consorting with wicked angels and there were abnormal giants in the land. At such an early stage of earthly tenure these things could not be allowed to go unchecked. God repented that he had ever created humankind. It is likely that at Sodom people had become infected with diseases. If these were not eradicated drastically the human race could have been wiped out or seriously impaired. Consider the effect of AIDS on our present generation. God had to ensure that there would be life on earth so that he could intervene in history by the incarnation of His Son. It should not be overlooked that Satan played a role in these cases of annihilation by leading people into sin because he had a vested

interest in preventing the incarnation from taking place. In his famous prayer, Jesus asked that people should be delivered from evil and prayed that the kingdom would come.

(11) Falling away. As further evidence as to why the kingdom is hidden from strangers it is useful to view these obstructions from another angle. A study conducted in America looked at reasons why Catholics fell from their faith. It was found that only a small number did so on account of doctrinal dissatisfaction (less than one tenth). Twice that number left because they openly preferred more worldly life styles. The survey disclosed nothing about disillusionment brought about by the behaviour of priests. The remainder had either lost interest in spiritual things or had only remained in the church as long as they were under parental influence. Detail about disaffiliation from other mainstream churches is not so clear but there was evidence that the god of this world figured quite prominently. Older people move out of churches that adopt new and unfamiliar patterns of worship. In the case of modern brainwashing sects that lure strangers away from seeking the truth there were large drop-outs as a result of deprogramming. Sadly a large core remained in bondage to those who were making merchandise of them.

(12) Light and darkness. It is good to be able to end positively in what has been largely a negative chapter. Encouraging news for strangers is that with the coming of the Son of God into history the kingdom of God began to act in powerful ways against evil. Jesus said that if he cast out demons by the finger of God the kingdom had come upon people. Jesus bound Satan and plundered his house. Spiritually blind eyes were opened by the same one who said to the physically blind man, "receive your sight." Satan cannot open the eyes of physically blind people but he does blind the spiritual eyes of unbelievers so that they do not see the light of the glorious gospel of Christ, who is the express image of the Father. That is why the spiritually blinded mocked at the cross by crying out that Jesus had saved others but could not save himself. The same blindness led people to doubt the final consummation of the kingdom. Peter wrote, "In the last days scoffers will come asking where is the kingdom he promised? Ever since our fathers died, everything goes on just as it has since the beginning of creation." Jesus said of the end that it would be just as it was in the days of Noah. People would be going about their normal lives quite oblivious of what was to overtake them. They would be marrying and giving in marriage. This blindness produces apathy. This is borne out by changes of heart in times of crisis. At the commencement of major wars national churches begin to fill only to empty again once peace returns. True

believers were told by Jesus to so let their light shine before men that their good deeds would be seen and his Father in heaven would be glorified.

This chapter has sought to show that news of the true kingdom can be hidden amongst masses of conflicting information, by people being preoccupied with worldly affairs, by ever proliferating denominationalism, bad witness caused by unacceptable conduct, and a lack of positive witness on the part of citizens, by extra-biblical doctrines and the propagation of wrong teachings, but that the dynamic kingdom is impacting daily on forces of darkness and bringing people into his glorious light. Paul saw the glory of the cross. He wrote, "God forbid that I should glory, save in the cross of Our Lord Jesus Christ by whom the world is crucified to me. And I unto the world."

4 | Hidden From Citizens

"Now we have not received the spirit of the world, but the spirit which is of God;
That we might know the things that are freely given to us of God."
1 *Corinthians 2:12*

The reality of a present dynamic kingdom can be hidden from its actual citizens for a number of reasons.

*(1) **Insufficient recognition.*** There are a number of spiritual blessings that are of great moment but unfortunately possession of the kingdom is not ranked as of equal importance to any of these. Such more highly regarded blessings include new life in Christ, resultant fruit bearing, power of The Holy Spirit, assurance, and two-way communion (i.e. praying to God and having the Lord as intercessor). Now supposing that all these blessings were wheels in God's chronometer and that his finger was the mainspring, then the timepiece as a whole would be his present kingdom. The most important part of this analogy is that of being so taken with the mechanism that the purpose and practicality of the chronometer is overlooked. This demonstrates that all these blessings collectively are the mechanism that bring peace, and righteousness and joy in The Holy Spirit which is in fact what the kingdom is, and that this truth is not, generally, acted upon with sufficient conviction.

*(2) **Confusion of doctrines.*** The simplicity that there is in Christ and the gospel and the kingdom of God has been neglected over the centuries because churches have been preoccupied with the formulation of various man made doctrines. Jesus was aware of this problem in Jewry. The Apostle Paul too knew that is was a problem in the early church. He wrote, "God is not the author of confusion but peace." He

also asked, "What came the word of God out from you? Or came it unto you only?" Nobody has a right to water down God's word or to add to it. Those who do so hide the kingdom. It is not the purpose here to comment on who has been right and wrong over the centuries because that would take many pages. It will be sufficient to draw attention to the perplexing plethora of factions that has distracted people from embracing the news of the kingdom. Augustine of Hippo 354-430 defended faith against ancient religions, philosophies and heresies. Interestingly, his best known writing was the *City of God* (Civitas Dei). He developed a doctrine of grace and assurance but his teachings have proved to be both enlightening in some aspects and divisive in others. He inspired Luther (who was also affected by the Epistle to the Galatians) to rediscover justification by faith but he also set a number of hares running including Calvinism. Luther was convinced that man did not have free will and this eventually led modernist Protestants into a form of Pantheism. It is said of these that they began by telling people to open and read the Gospels and to go to Christ, and then finished by shrouding the gospels in a winding sheet of sceptical scholasticism and erasing the features of the Redeemer. On the other hand, Ignatius Loyola, who encouraged spiritual exercises, took justification quite differently. He taught free will but encouraged a form of antinomianism, salvation by works, and opposed assurance. His views were magnified by Alphonsus Ligouri, who is the father of modern Catholicism. He influenced doctrines of worshipping the sacred hearts of Jesus, Mary and Joseph, Papal Infallibility, and the immaculate conception of Mary. The greatest bone of contention was on the subject of whether bishops drew their authority from God or the Pope. Tribute must be paid to Jansenists and Gallicans who attempted to get back to the simplicity of the bible. The latter were defeated within Catholicism by the Ultramontanes who were supported by Jesuit casuists. A disservice is done to the kingdom by sophists because these people write with tongue in cheek. G. K. Chesterton was a brilliant writer and poet. He was given the Papal accolade of Defender of the Faith. His argument was that although the church had reeled drunkenly, swaying from the right to the left over the centuries, this was misleading. If one observed the overall direction it followed this would be in a straight line. Cardinal Newman made no such excuse. Rule XIII of the Jesuits was – "we ought always be ready to believe that which seems white is black if the hierarchical church defines it."

(3) The Spoilers. There are a number of ways in which citizens can have kingdom teaching spoiled for them. Firstly, there can be pretence to gifts of The Holy Spirit. The bible gives instances in which God speaks to people, but it also condemns others for saying, "The Lord

says" when he has not. Jesus was very clear in saying that there has to be a waiting on God. What is happening too often today is an appropriation by individuals of what is written without any genuine operation of the spirit. This results in sincere people quenching the real power of the kingdom. Secondly, there are teachings that can present the kingdom in a false light. These are exaggerated but often populist. A false prosperity gospel is being spread amongst the gullible. Give to God and reap a hundredfold return. This warps a promise by Peter about an eternal reward to those who give up everything in this age, by applying it in an unbalanced way to fill the pockets of those who follow after the doctrines of Balaam. People might do a lottery or back race horses and lose. They think the prosperity gospel offers better odds. They are often from very poor countries. They end up disillusioned. Then there are those who are unbalanced about divine healing. Jesus always put less emphasis on healing and usually told benefactors not to tell others. His actions were compassionate and his intentions were to glorify his Father. There are encouragements in the scriptures to pray for one another for healing and there is a gift of healing. Balanced study shows that healing is neither a must nor a norm. Everybody deteriorates with age and dies. It is the penalty of Adam. Jesus was attacked when he pointed out that there were many lepers in the day of Naaman the Syrian but that only he was healed. He also indicated that signs would not be given to a wicked and false generation except for the sign of Jonah. He meant that just as Jonah was in the whale for three days, Jesus, as Son of Man was three days in the earth before rising from the dead. Jesus meets many personal needs of his fold but the lower the key it is placed in the better it is for the work of the kingdom. The priorities are to call for repentance (a changed life) and entry into the kingdom by a second (spiritual) birth. What can be more exploitative and ruinous than calling a public healing meeting at which nobody is really healed?

(4) A form of godliness. There are those who have a form of godliness but deny the power thereof. These liberal teachers put a damper on citizens and strangers alike. In fact disbelief in the literal resurrection of the Lord and in the miraculous is more damaging to faith than the attacks of hard-nosed atheists. Paul said that if Christ has not been raised our preaching is useless. A highly respected bible scholar passed off Christ's walking on the water as an illusion. Christ, he said, remained on the shore and his followers' boat was tossing up and down. It only appeared that Jesus actually walked on the water. Any seafarer will be able to relate such a situation to when two ships are passing in rough water. It can look as though the other boat is up in the sky. The critic failed to comment on the calming of the squall because he could not

invent an alternative explanation! The difference between "what is that" and "what isn't that" in relation to the power of the spirit is discussed elsewhere.

(5) The unequal yoke. Citizens can be severely tested by being obliged to mix with people whose backgrounds are ungodly. Having to leave a Christian household on being conscripted in the armed forces is challenging, as is going into full-time education in another town or county. Even Christian unions are not always safe havens. These can be joined by proselytes peddling extremist doctrines that bring the standing of others into doubt. It can be challenging to citizens if they nail their colours to the mast in the workplace. It presents an opportunity for those who are strong to act as salt. It is not uncommon for citizens to adopt a role of secret disciple although they have not been given a spirit of timidity but of power and of love and a sound mind. Jesus said, "A city set on a hill cannot be hidden. So let your light shine before people that they may see your good works and glorify my Father who is in heaven." Citizens are expected to be active in generating the work of the kingdom. The Holy Spirit in executing the counsels of God operates both directly in the work and indirectly by empowering individuals. As with Timothy there is a constant need to stir up both ministry and spiritual gifts of the Holy Spirit.

(6) Holding the truth in unrighteousness. This is particularly damaging to faith when this trait is found in leaders. If these are exposed as adulterers, abusers, or embezzlers, or unbelievers in heart, the shock to others can result in love growing cold. Other traits are damaging even if less serious, for instance Paul wrote about busybodies and backbiters, whose behaviour did not edify the church. In smaller churches one can encounter nepotism and striving after control. The autocratic attitude of priests has been replaced by greater friendliness, no doubt partly because people are more independent, and partly because this generation is less favourably inclined towards regimentation as a whole.

(7) Apathy. Many church members are happy to occupy pews. With an increase in worship, music, drama, and anecdotal sermons, which are acceptable features at the present time, there is insufficient room for expository ministry of the word. Although bible study is catered for in house groups, these can be led by people without teaching gifts. Some groups are weighted towards equal participation, gaining much in fellowship but not so much in scriptural knowledge. Disaffection is more likely to be on account of style of worship, whether exuberant or reverent, noisy or quiet, rather than what is or is not being taught. In the Book of Malachi it is said that those who feared the Lord spoke often one to another . . . and a book of remembrance was written before him that thought upon his name. This means that they spoke

about the Lord. Is it the norm today before and after services and meetings? If it was, surely there would be an increased appetite for the things of his kingdom. The percentage of populations attending prayer meetings is disappointingly low. Alpha courses for non-church goers are achieving good responses. Perhaps there should be Omega courses for church goers?

(8) The media, publishing and censorship. With the advent of television and more recently satellite God channels citizens are even more distracted from the gospel of the kingdom. The time given to Christian services by the BBC and ITV is minimal. It is regrettable that a state funded (by taxation) channel does not cater proportionately for sections of society. There are many Christians in homes who have not access to a service. Because programmers have control even when there is a service like Songs of Praise – hymns are sanitised and professional singers and actors introduced to the exclusion of members of the church involved. The God Channels do have a few good programmes but these are eclipsed by a very large number of presentations that sour both balanced citizens and discerning strangers alike. Enlightenment came in the midst of darkness with the invention of the printing press so the kingdom owed much to publishing and still does. Good publishers are user friendly. However, there are others that will reject whatever does not follow a particular doctrinal line and so suppress freedom of thought. The reverse is true of some Christian bookshops which stock any work that will sell and so disseminate outlandish books full of exaggeration and spurious claims. An excuse given is that there is not time to read through the stock. This is a disservice.

(9) Fanaticism. Citizens can unwittingly get caught up in meetings conducted by fanatical sects whose leaders are deluded or charlatans. One thing is sure. That is vast sums of money are involved. Because of the techniques of extortion employed individuals pay far more to these sects than they would to their own churches. There are claims to revelations, unsubstantiated "healings", and false prophecies. Meetings are marked by lack of self control and hysteria. Many are drawn away from moderate churches. People are so taken up with looking for signs and wonders that the proper business of the kingdom is forgotten. Mature citizens do nevertheless exercise judgment on what is being preached. However, it takes a very discerning person to realise what is not being said. There are many rational people, especially in countries with long religious histories, who regard too much involvement in spiritual matters as leading to undesirable fanaticism. This is an important example of how fanaticism can detract from genuine spirituality.

(10) Wrong priorities. Dedicated citizens put their hands to the plough and do not look back. Those who do look back are adjudged by

the Lord as not being fit for this work. There are those who are anxious to serve the kingdom while they are alive. There are others who find greater pleasure in entertainments, secular pastimes, and an excessive number of holidays. While striking a balance is not unreasonable it is important to realise that the gifted should not be neglecting to use what God has bestowed upon them. Everybody should reflect that a unilateral promulgation of news of a present kingdom motivated by God's finger would awaken those from whom it is hidden and there would be many new entrants who were formerly just religious.

5 | Entering the Kingdom

"Behold, I stand at the door and knock: if any man hear my voice, and open the door, I will come in to him, and will sup with him, and he with me." *Revelation 3:20*

There are scriptures that support the existence of a present earthly divine kingdom and a gift of new life that admits people into citizenship of it. Paul wrote that God has delivered us out of darkness and has translated us into the kingdom of his dear Son. Peter wrote that God has called us out of darkness into his marvellous light. When a religious leader came to see Jesus by night he was told: except a man is born again he cannot see the kingdom of God. "The wind blows where it will and one does not know where it comes from or where it goes to. So is everyone who is born of the spirit." Jesus gave Peter the keys of the kingdom by saying, "You are Peter and upon this rock I will build my church and the gates of hell shall not prevail against it."

Commentators spend a lot of time speculating as to who or what the rock was. It is more important to state what was to be achieved. This was the impartation of a new nature to be possessed by members of Christ's true church. In that Peter preached this, he had in effect the keys that opened the kingdom. On the day of Pentecost Peter preached the full gospel of the kingdom. The people responded by asking what they needed to do. Peter said they must repent (turn about) and be baptized in the name of Jesus for the remission of sins and they would receive the gift of the Holy Spirit. Later he wrote that God had begotten them again to a lively hope by the resurrection of Jesus Christ from the dead to an inheritance that did not pass away. Peter started to sow the word of God. They were born again of an incorruptible seed by the living and enduring word.

Faith comes by hearing the word. It is also useful to note that the bible uses the symbol of water to depict the word. Otherwise one can be led into thinking that there is something efficacious in water itself. Water merely symbolises baptism which is not just the removal of dirt from the body but a good conscience towards God. A new birth can be brought about in a number of ways. It is not every citizen that has had a Damascus Road experience. There is always the work of the Holy Spirit and the Father draws people by this means. There must also be a responsible reception. This can be by calling on the Lord, by confessing the Lord, or growing up into Him. It also involves turning from an old life and hungering and thirsting after righteousness. There will be evidence of new birth. Otherwise an entrance into the kingdom does not take place. There should be joy and happiness but to what degree is a personal thing.

There must be evidence of a changed life for the better. In the event of failure there will be contriteness whereas before there would have been indifference or even pleasure. New entrants are aware that they have changed but it is difficult for strangers to imagine what Christian assurance is like as they have not experienced it. Poor and uneducated believers found it difficult to explain their conversions from the scriptures. However, knowing what work God had wrought in their hearts they would say, "It's better felt than tellt." The Spirit had borne witness with their spirit that they were children of God. Another heartening fact is that this newly acquired life is inviolable. Because the gates of hell will not prevail against Christ's church it follows that no member of it will ever go through them. As a reminder, Christ's church consists only of those who are in the kingdom. This inviolability is underlined by many scriptures. Who shall separate us from the love of Christ? Shall trouble, hardship, persecution or famine or nakedness or damage or sword? No, in all these things we are more than conquerors through him who loved us . . . neither height nor depth nor anything in creation (or any other person = AV) will be able to separate us from the love of God that is in Christ Jesus our Lord. Some denominations are quick to apply the "gates of hell" scripture to themselves. That they are wrong is borne out convincingly by many unfortunate events in church history. God brought individuals into an inviolable position in His kingdom as a master solution to a long standing problem. People had failed, firstly in the Garden of Eden, and secondly under the law. He also needed to separate them from the world. They were remade anew. By being partakers of His nature they were equipped to fight against their Adamic or sinful nature. The waters of baptism are a symbol of separation as Peter pointed out when he explained that as well as being saved physically in the Ark, Noah and his family were saved in a spiritual sense as well by the waters separating them from the world.

But what then about the mainspring by which entry into the kingdom was brought about in the first place? God is holy and just and also merciful. Justice and holiness are not principles that can be overlooked in the long run. Transgression could only be winked at for a season. As things stood Adam and Eve could not be exonerated when Satan and his wicked angels had not been. God found not only a way of forgiveness but also a means of providing eternal security. The world was spiritually dead in transgression and sin. Justice demanded propitiation and this would be offered by a perfect substitute. Jesus was not born by the will of the flesh but of a virgin by the power of the Holy Spirit. Because of this he was the first born of a new creation. In our parlance, he was a prototype and would thereafter bring many to glory. He was God in a body that was in the likeness of sinful flesh. As God, by taking on our human nature, he was able to impart divine nature to us. By obedience He died in our place the just for the unjust to bring us to God. This satisfied justice and yielded mercy and peace.

There are some warped views to the effect that the bible portrays God as an abuser for allowing his Son to suffer. Apart from the fact that Christ's death was an overwhelming act of love by which mankind was offered an escape from a second or spiritual death and a possibility of receiving the gift of eternal life there were also trophies for his Son. He was raised from the dead and declared to be the Son of God with power. The glory he had with the Father from the beginning was restored to him. It should also be understood that the nature of God is to desire fellowship with his created beings. This is why Jesus gave this illustration: except a corn of wheat fall into the ground and die it abideth alone but if it dies it brings forth much fruit. Just consider how far the powers of good have triumphed over all the powers of darkness since Satan revolted and mankind became disobedient. What a haven has been provided in this present invisible kingdom and what a hope there now is of eternal glory! In a longitudinal sense there are many unborn who will enter the kingdom. The proportion of such persons to overall populations will depend upon the effectiveness of witness. Whether there will be a general conversion of the nations will be a matter for further discussion.

6 | Life in the Kingdom

*"A spleeny Lutheran; . . . And 'tis a kind of good deed to say well;
And yet, words are no deeds." Shakespeare, King Henry VIII, Part II*

Jesus told a story about two sons. Their father asked them both to go out and work in his vineyard. The first one said he would go and did not. The second said he would not go but did. This is a powerful illustration of the difference between empty profession and effective service in God's vineyard. It is like this when some hymns are sung with fervour such as, "Wherever he leads I go," and there is no actual response. Life in the Kingdom is an unmerited gift. However, it is unthinkable that such a precious and costly endowment would not evoke in recipients a fervent desire to manifest the fruits of righteousness that are by Christ Jesus and to be found in his perfect will. The love of Christ should constrain us to be his ambassadors by preaching (or gossiping the gospel) and living a good life. It is all too easy to accept the precious gift and to take a selfish view that one's personal safety is all that matters. Let us settle down go to church and enjoy a quiet and contented life.

Of course, not everyone is called specially. There were occasions when Jesus expected people to stay in secular occupations. Paul also told people to remain in the calling to which they belonged. To a new creature deeds should become important. The work of the temporal kingdom suffers when citizens get stuck on being justified by faith. There was this terrible tension between the views of the Roman and Reformed traditions. The former recognised the correct morality of a life of good works but saw these as a means to salvation in themselves. This was the error of religious Jews who being ignorant of God's righteousness went about to establish their own righteousness. The latter soft peddled on good works partly because of disagreement with the Roman tradition, partly because of the surprise and joy of Luther's rediscovery of a scriptural truth, and partly because of being obsessed with it. Luther even challenged the authenticity of the Epistle of James because it stated faith without works was dead. Where is your faith? Show us your works! If there had been more debate this controversy could have been avoided and the fact of a present kingdom established more positively. The biblical truth is that it is by grace citizens were saved through faith, and that not of themselves, it was the gift of God, not of works lest anyone should boast. But they have been created unto good works which God has foreordained they should walk in.

It is controversial to what extent, or if at all, citizens will lose out because of apathy or backsliding. At the very least they will be ashamed, or lose rewards. There is no doubt that they incur disapproval because Jesus said that if anyone put a hand to the plough and looked back they were not fit for the kingdom of God. Once the reality of a present kingdom is recognised scriptures like this (and there are many others) come to greater life. This particular scripture underlines the importance of life in the present kingdom once it is entered. A good tree should bring forth good fruit: not none, or rotten.

The person who would win others should be wise. If within a household an opportunity arises to explain the gospel to the young it is unprofitable if seemingly mature older believers engage in a heated debate about whether the church will go through the great tribulation or not. On the subject of tribulation generally it is debatable whether citizens lose out in this life on account of spiritual failure. It is possible that sin might result in early death but there is no positive evidence of this because so many known good people are called home early whereas the ungodly prosper in the world and increase in riches. When Jesus was tempted by the prince of this world with all its kingdoms it is obvious that this was regarded as within his competence.

What is certain is that citizens can be chastened and pruned with a view to their own good. No chastening for the present seems joyous but grievous: nevertheless afterwards it yields the peaceable fruits of righteousness to those who are exercised thereby. This implies a need to recognize the chastening. An understanding of this helps citizens to appreciate the character of the present kingdom. If the wicked prosper they will face judgment eventually as was shown to the Psalmist when he went into the sanctuary. Life in the kingdom is about conflict and this is on two fronts. First of all citizens experience a conflict within between their old sinful nature and the new nature imparted by the Spirit. Well did Charles Wesley pen: Thy gracious nature Lord impart. Come quickly from above. Write Thy new name upon my heart, Thy new best name of love. The sinful nature desires what is contrary to the Spirit, and the Spirit what is contrary to the sinful nature. They are in conflict one with each other so that you do not do what you want. Jesus said it was necessary to take up the cross daily and to follow him. He meant that in order to deny oneself the sinful nature had to be reckoned dead as crucified with him. This is a reckoning by faith as opposed to a reality. There will be no absolute perfection in this life. There must be no confidence in the old nature. There is power in the new nature to enable citizens to walk in newness of life because of the Spirit of Christ. This is invincible. Whosoever is born of God does not sin; but he that is begotten of God keeps himself, and that wicked one touches him not. Failure of the sinful nature results in Jesus acting as an advocate with the Father. So much for the inner struggle!

As to taking the fight on to enemy territory the overcoming citizen will walk in the Spirit. They who are led by the Spirit are the children of God. God gives incomparably great power to citizens. It is so great as to be comparable to that which he exerted in Christ when he raised him from the dead and seated him in heavenly realms far above all rule and authority, power and dominion. Then ministry gifts are set in the church not only to shepherd but also to prepare God's people for works of service. In order to be strong in the Lord and his mighty power citizens must put on the whole armour of God so that they can take a stand against the devil's schemes. The struggle is against the powers of this dark world and against the spiritual forces of evil in heavenly realms. The armour is truth, righteousness, the word, faith, and the new life, and prayer in the Spirit.

In the kingdom parable of the hidden treasure and the pearl of great price the lesson was that one should go to great lengths to find the Kingdom of God even to the extent of giving up everything else. Another aspect was brought home by an advertisement which asked whether people were wearing their jewellery or just insuring it. People are afraid to wear expensive jewellery because of robbers. They put it in a bank deposit and often wear imitations. Now as to the parable, the treasure would not have remained hidden once the field had been acquired. Neither would the pearl have been left in a vault. The analogy is that citizens should not hide the Kingdom of God by keeping its secrets to themselves. The bible states that the mystery hidden from the ages is Christ in us, the hope of glory. We have this treasure in earthen vessels in order that the excellency of the power is of God and not of us. The earthen vessels relates to our human bodies and their propensities. If we do not let our light shine before others it is no better than leaving a treasure buried permanently in a field.

A little while ago a young mother who was not a Christian picked up a Gideon Youth Testament her son was given at school. She was accustomed to reading the end of books first, particularly if these were "who dunnits", to find out quickly who had. Better go straight to the end – The Book of Revelation. Dramatically affected by its lurid apocalyptic prose she sallied forth and told everybody on the way what was to come upon them. The living daylights were scared out of not a few people. Some of her interpretations were rather novel. Now ought not the tremendous news of the kingdom be spread abroad with the same enthusiasm?

There are some A-millenarians who believe that a present second birth is in fact "the second resurrection". There are some uplifting scriptures that have led to such a supposition. For instance, Jesus said: "I tell you the truth, whoever hears my word and believes on him that sent me will not be condemned: he has crossed over from death to life. I tell you the truth, a time is coming and has now come when the dead will hear the voice of the Son of God and live." A close study of these words will show that Jesus was talking about those who were spiritually dead being raised to newness of life. But they are still in Adam and their bodies will not be raised until the second coming of The Lord. It is marvellous in our eyes that as citizens we reign in life through Christ Jesus. Our lives are hidden with Christ in God: the Christ who is mediatorial king whom the heavens hold until the restoration of all things. But that has nothing to do with the glory that will be revealed when the redeemed are conformed completely to the image of Christ.

Life in the kingdom is not only doing and listening. There is provision and companionship. When Paul went into Macedonia he was tired out. Both he and Titus were refreshed by the concern people had for him and by the comfort they afforded them. Paul likened the kingdom ("The body of Christ") to different citizens representing different parts of a body – one being a foot and another a hand. Just as a body is framed together and all of its parts necessary in which the hand cannot say to the foot, "I can do without you" – citizens are bound together. When one rejoices everyone rejoices as well. When one suffers all suffer. Faith without works is dead. If a brother or a sister is without clothes or food and are given best wishes for keeping warm and being fed and nothing is done about their physical need, what good is it? Then again loves flows out of the kingdom. This can be substantiated by what Jesus will say when he returns. There will be those who fed the hungry, gave drink to the thirsty, clothed the needy, housed the homeless, nursed the sick and visited the imprisoned. He will say to them that as much as they did it to the least these strangers they did it to him.

That is why the work of The Salvation Army, and Medicine Sans Frontiere must be very close to God's heart. Lower profile bodies like the Catholic Society of St. Vincent de Paul, and the independent Missionary Aviation Fellowship also deserve a mention. Life in the kingdom is largely a matter of faith and this is the subject of the following chapter.

7 | Faith and the Kingdom

"Now faith is the substance of things hoped for, the evidence of things not seen." *Hebrews 11:1*

He that comes to God must believe that He is and that He is a rewarder of those who diligently seek Him. God said to Abraham, "I am your exceedingly great reward." This was sufficient in itself. Implicit belief and trust in God that He will do right is a keystone of faith. The following is an obiter dictum (aside) to demonstrate that belief in God rather than people will lead to enlightenment. There is a purpose in this because it is a principle that will be applied to good effect when difficulties that have beset expositors in the past are considered. It will be most helpful when the end days are discussed in Part Two.

The prophet Daniel foretold, "seventy weeks are determined upon thy holy city, to finish transgression, and to make an end of sins, and to make reconciliation for iniquity, and to bring in everlasting righteousness, and seal up the vision and prophecy and to anoint the most holy. Know, therefore, and understand that from the going forth of the commandment to restore and rebuild Jerusalem unto the Messiah the Prince shall be seven weeks and three score and two weeks: the street shall be built again, and the wall even in troublous times. And after three score and two weeks shall Messiah be cut off but not for himself." This prophecy pinpoints the length of time between an order to rebuild Jerusalem and the cutting off of the Messiah. It is generally agreed "seventy weeks" means seventy times seven or 490 years. Numerous scholars tried to fit this prediction into subsequent history. They all took the order to relate to a decree by Cyrus the King for the edification of the temple. As none of the computations fitted it was decided that the prophecy could not relate to the coming of the Messiah at all. It must relate to other times and events. The less generous suggested the prophecy was flawed and this was a drawback to faith. However, John Owen the English theologian decided to believe God. He worked backwards from the time of Christ's cutting off and arrived at a starting point that exactly fitted to a more obvious decree. The decree of Cyrus was not for the rebuilding of Jerusalem but the temple only and in the event the people did little more than build a mere fabric. The decree Owen discovered was by King Longimanus with which he sent Ezra to Jerusalem. It was the most useful to the people out of all the royal orders. It was not just a proclamation like that of Cyrus but a law made by the king and his seven counsellors. It not only involved the temple but also setting up a civic government and the building of the city mentioned by the angel Gabriel. This extra-ordinary discovery fulfilled the prophecy. The time between the issue of Longimanus's order and the death of Jesus was exactly as foretold by Daniel. This salutary and faith inspiring lesson is a reminder not to take things at face value.

When the people asked whether anything good could come out of Nazareth they were completely on the wrong track not knowing that Jesus was born in Bethlehem of Judah. Herod's advisers did not make the same mistake, being aware of the prophecy that out of Bethlehem would come forth a leader that would save Israel. Disciples did not believe the women who had seen the empty tomb and followers did not believe Peter had been miraculously freed from prison thinking it was his ghost knocking on the door. Paul said he knew in whom he had believed and was persuaded He could keep that which was delivered to him until the last day.

It is less easy to believe others when they testify about the faithfulness of God in their lives so there is a greater value to the individual recipient than to those to whom it is passed on. It is also reasonable to assess and to test unconfirmed accounts. Sometimes there is concrete evidence like when a man said, "all I know is that I was blind and now I see." Many knew he had been blind since birth. Some then queried the source of the healing.

Lack of faith leads to disobedience, which has its roots in thinking that following of commandments does not matter. The children of Israel were instructed to leave a field fallow every seven years. The instruction was disregarded from David's day until the captivity in Babylon – a period of 490 years. Apart from disobeying God over what he regarded as his special territory (Canaan) it was bad agricultural policy. Until fertilizers were introduced a rotation of crops was a successful practice in most countries for centuries. It would have been healthier to stay with God's instruction despite Malthusian theories. The captivity in Babylon lasted for seventy years and fields, therefore, laid fallow for the exact length of time over which God was disobeyed: that is one year in seven for 49o years – a total of seventy years! This was remarkable and faith inspiring. An injunction to rest on the seventh day was sound physical reasoning, as well as to devote it to worship.

Today there is a great deal of burning the candle at both ends instead of following God's commandments. Sometimes a promise or a direction is taken from scripture in faith and proves to be true. This is opposed to practices of appropriating scriptures in a manner that is not warranted such as passing a promise box around like a box of chocolates. A Christian husband and wife ran an orphanage. Questionably, but out of love, they adopted a policy of never refusing to admit a child. The authorities disapproved of the cramped conditions that were contrary to expected standards. One day all the children were removed in buses and the orphanage was left empty. Bereft of the children they went to church looking for guidance from scripture. It was a strange church to them in America. As they entered a reading was being given from the 49th Chapter of Isaiah. Although these verses relate to the restoration of the true Israel it was taken that these applied to their situation, particularly that the children they would have after they had lost the others would be brought to them on peoples' shoulders and they would be received by kings. The couple went to Africa with two young helpers in complete faith and set up an orphanage in a remote part of the bush.

An expatriate colonial civil servant and his wife had begun to look after an African boy who was being badly treated. After sometime they were due to leave the country. Because it was better to leave the boy in his own environment and culture an alternative source of caring was sought. They drove some miles north to a missionary station and it so happened that the missionary had returned that very moment from the orphanage already referred to. It was very many miles farther north. He was certain that because of the non-refusal policy the boy could be placed but it was stipulated that the expatriates must take him in themselves. This was seemingly impossible because as a senior person the civil servant was tied to the capital city and in any case had entered into an involved takeover of duties. On going into his office the next morning he learned that due to serious rioting in a northern town and the disruption of communications a senior person was needed to carry confidential dispatches. Fortuitously, he and his wife and the boy set off the very next day. After delivering the dispatches they struck further north still and stayed in a village overnight with an elderly lady missionary who was the only white person in that area and held in high regard. Next day proper roads terminated and sometimes it was a matter of steering between trees. Coming into a clearing three white women were seen working in a field. Two were wearing trousers, which was unusual for missionaries in those times. Inside a building that was the orphanage were seventy young children all dressed for bed. There was cocoa on a stove. The man had a squeeze box and they were singing, "I will sing of the mercies of the Lord forever." The boy was taken in and some money handed over. There was also a trunk full of things no longer required in colder climates such as summer dresses. It transpired that money had run out that morning. The reason for trouser wearing was that they only had one dress out of the wash so the extra dresses were filling an immediate need. Most of remaining contents of the trunk turned out to be necessities such as cutlery and even a table tennis set. The work went through a period of enmity from witchdoctors. A beheaded animal was thrown on to the doorstep. The kings and queens of that northern territory came to respect and admire the faith and dedication of the missionaries and began to support them in a substantial way. A much larger orphanage was built even further north.

There are innumerable instances of the outworking of faith each day that are known only to those who exercise it and inner groups who are familiar with the circumstances. These are also known and noted by the king. Cases that are or were in the public knowledge will form the subject of the following chapter.

A big test of faith is in trusting that God will provide when all else fails. God is not running a benefits system for dodgers. Those that won't work shouldn't eat. Paul worked as a tent maker to ease the church budget despite the fact that he was entitled to payment on the basis that the ox that treads out the corn should not be muzzled. William Romaine was a godly clergyman whose parish was in London. A book about him was called *A Life of Faith*. The church deeds stipulated that a minister was to be elected annually. In alternate years power of appointment switched from parish council to the members of the congregation. Because he preached the gospel and was not ritualistic he was out of work every other year. He lived by faith during the periods of unemployment. When he departed this life for heavenly territory over 2,000 people attended his funeral.

Some years ago a married couple with a teenage son, an elderly parent and a Labrador dog needed to set up home afresh after having served abroad. To make ends meet the wife supplemented the husband's income by working for The National Assistance Board at a salary of £550 a year. It was really too much of a strain for the wife to run the household and to work as well. They came upon the reading: "Consider the lilies of the field. They toil not neither do they spin, but Solomon in all his glory was not arrayed as one of these. And if God does so clothe the grass of the field that today is and tomorrow is cast into the oven, how much more will he not clothe you, O you of little faith?" On the basis of this scripture the couple agreed that the wife should resign. A few days later the husband received a pay increase of £550 a year.

In the twentieth century The Scripture Gift Mission had a policy of Jehovah Jireh (The Lord will provide). It never asked for money. It was faced by a shortfall of £150,000 in its finances. The Board of Management got down on its knees and prayed. The next day it was asked by the Laing trust if it could use £150,000. There are not many who would sell all that they have and give it to the poor. C. T. Studd, a member of the middle upper-classes, and a successful test cricketer, did just this. He joined the China Inland Mission and was one of the famous five. His friends disapproved because it seemed that his children would miss out on a private education. However, under conviction of the Holy Spirit, they felt obliged to stump up!

God does not believe in credit, or in usury for that matter, although in secular life most people are obliged to use it. God must never be put to the test. If for example, it is decided to build a new church and it is in the permissive will of God, he will enable his people either to save up for it or to pay outright. The subject of faith will spill over into the next chapter.

Faith can be of varying intensities but one can start with little. Jesus likened this to a grain of mustard seed and he is the one who can help unbelief. Jesus was touched whenever he encountered a strong faith in people. When a Roman centurion wanted Jesus to heal a sick servant Jesus proposed to accompany him to his house whereupon the centurion said he was not worthy to enter under his roof. "Say but the word and my servant for Jesus will be healed." Jesus said he had not found so great a faith in all Israel. The servant was healed from that hour.

In Matthew's gospel it is recorded that Jesus said those of genuine faith will take their places at the feast with Abraham, Isaac and Jacob in the kingdom of heaven. But the subjects of the kingdom (i.e. presumably historic Israel) will be thrown outside into the darkness. This last sentence would prove difficult if it wasn't for Luke's rendering in which the contrast is between the righteous and the wicked. The Jews were called children of the kingdom because it was to them that the Messiah was promised. Jesus was contrasting wicked and righteous Jews. The righteous had repented and been born again of the Spirit. They would then go on to perform works of righteousness that God foreordained they should walk in. These works, if inspired by appropriate faith, can result in the moving of mountains. It must not waver or one will not receive anything from the Lord.

What has been examined is ordinary faith. There is also a special spiritual gift of faith that can be bestowed as God wills. It seems to have been given and exercised in the past in cases of necessity. George Muller is known to have exercised unusual faith with remarkable results. He was able to run orphanages for hundreds of children with no obvious means of support. Hudson Taylor was under a death sentence in a remote part of China. There would be no execution but the population was forbidden to supply him with food. He was praying in a shuttered upstairs room. There was a knocking on the outside of the shutter. On investigation it proved to be a man up a ladder with a parcel of food.

Personal faith is something that will endure. Citizens will persevere to the end: more importantly this will be because they will live by the faith of Jesus. Paul wrote that the life he lived in the body he lived by the faith of the son of God who loved him and gave himself up to death for him. Having faith in God means being strengthened by all his statements and promises and acting upon them. Faith in a present Kingdom is a dimension that many real believers lack because they are unaware of it. This means that although they are blessed with the certainty of salvation and the knowledge that the heavenly territory will receive them on death they are not enthused by the existing spiritual Kingdom and its dynamic working in the here and now.

Then there is a difference between faith in God and faith in a faith. It is a tragedy that so many people are taken up with a faith. This is born out by the incident just after Jesus met the centurion. He met those who saw Jewish faith as an end in itself. Their basis was misplaced. Of course, Kingdom faith is an ongoing one as opposed to a one off profession. Jesus made this clear in the parable of the sower. It was only the seed that fell into good soil that endured. The complex problem of the eternal destiny of those that the other soils represent will be discussed in a later chapter. The Bible teaches that there will be a first resurrection and a second resurrection. It might be a shock as to who is and who is not found at these.

This present life is certainly not a bed of roses and it is comforting that Paul calls all its problems comparatively small compared to the glory which will follow. In eternity the redeemed are not onlookers. They are involved and blessed. Not only will they see the glory of God. They will be glorified themselves. This glory will be revealed in them.

The next following chapter is about service. The Apostle James wrote, "Have you faith? Show me your works." One is not saved by works lest it result in boasting, but that works follow after genuine salvation is evidence that is has been brought about. Therefore, faith and works are inextricably bound together. It is important to understand that faith comes before works just as ordered in chapters 7 and 8.

8 | Servants of the Kingdom

"He that is down needs fear no fall,
He that is low no pride.
He that is humble ever shall
Have God to be his guide."
      John Bunyan

"Wherefore, we receiving a kingdom which cannot be moved, let us have grace, whereby we may serve God acceptably with reverence and godly fear: for our God is a consuming fire." *Hebrews 12, verses 28, 29*

As a start it will be good to stay with the continent of Africa. A great test of the real lives of people of faith is that their works do follow them. Very often there is a good report from without. When Nigeria gained its independence in 1960 it published a Special Independence Issue of its national magazine. The majority of its population, based mainly in the North, followed the Islamic faith whereas there were large numbers of Christians in the West and East as a result of missionary activities going back as far as 1840. The Muslims dominated the Federal as well as the northern government but had no qualms in paying tribute to the contribution of Christian missionaries in bringing people out of darkness and laying the foundations of education and culture.

The first missionaries had no drugs and prophylactics and most of them died within a few months or years. West Africa was known as the white mans' grave. They lived nearer to people than traders or officials, sharing their hardships, entering into their homes, learning their language, and sharing their burdens. Nobody is perfect but it was officially acknowledged that to the very limit of their resources they fed the hungry, cured the sick, freed the prisoner, sought mercy for the condemned, and comforted those in sorrow. Their reward was seen to be simply in doing of it and nothing else. A dedicated missionary, Henry Townsend, who served in Abeokuta, wrote in his journal: "This day renewed the vows I made to God two years since, to devote myself to His service in the missionary work with fresh entreaties for grace to enable me to go forth, with a single eye to His glory, and for full purpose of heart, not to return unless necessity compels me to do so, but to spend and be spent in His service; not seeking to please man but God."

A very heroic and devoted missionary was a woman, Mary Slessor, who worked chiefly among the up river tribes of Calabar province for 38 years, She fought to correct the notions of E K P E and witchcraft, to improve the status of women and children, to dispel fear and fight diseases. It is said that if you educate a man you educate an individual; if you educate a woman you educate a family. Women could enter into an intimacy of understanding with the mothers of Nigeria which would be most improper and unbecoming in a man. There were others at that time: "Mammy" Sutherland of Calabar, Anna Hinderer of Ibadan, and Sara Townsend of Abeokuta. In India there was Amy Carmichael of Donevar, and later Gladys Aylward and Isobel Kuhn of China.

There is not a lot of benefit in producing long lists of all the heroes of the Kingdom of God. It is a bit like genealogies in the Old Testament. These both have their place but are difficult for general reading. Present-day Christianity has shortcomings. Tradition has fostered sectarianism on one hand and syncretism on the other: ambition is as prominent as sacrifice. This is regrettable but, thankfully, there are many uplifting histories and current examples of servants of the kingdom that more than offset disappointing behaviour. It is reputed that all the apostles except Judas Iscariot ran a good race although details about some are not to be found in Acts of the Apostles. This book gives a very detailed account of Paul's missionary journeys which must have motivated numerous missionary aspirants. Other followers like Stephen, Luke, Mark and Barnabas (credited as an apostle) were stalwart servants of the kingdom. The eleventh chapter of Hebrews contains a record of Old Testament servants of God starting with Abel and finishing with Samuel and the prophets. Although these preceded the introduction of an earthly kingdom they were seen as a part of a great cloud of witnesses. The world was not worthy of them. They are enduring ensamples: none of them receiving what had been promised.

This thought is developed by Paul (the obvious writer of Hebrews) by reminding the citizens that they are able to fix their eyes upon Jesus who endured the cross and sat down at the right hand of God. Furthermore, they had received a kingdom that could not be moved (shaken = NIV). Therefore God is to be served with Godly reverence and fear. There is a further exhortation to strengthen feeble arms and weak knees with a view to becoming more active in the service of the kingdom. Two men who acted on these scriptures were John Wesley and George Whitfield. The actions of these two men greatly affected the lives of countless persons for the better, and assisted by the Holy Spirit drew them into the kingdom. Because they preached the true Gospel as opposed to ritualism they were shut out of state churches even though ordained by them. They went into the open air and crowds flocked to hear them. Like Jesus the poor heard them gladly and responded to their message. Righteousness and godly living could be observed in villages and towns where previously there had been surfeits of drunkenness and acts of lawlessness. Chapels sprung up everywhere.

Contemporary historians find it convenient in the face of secularism in the media to ignore this move of God's spirit especially in television series. It cannot be denied that the nation was spared troubles akin to the terrible revolution in France and that the revival of faith in England and Wales was historically of far greater consequence than things that are now being brought to the public attention about the period. Deceptively the presenters are extremely literate and otherwise credible individuals. Wesley and Whitfield did not agree about everything but it is an example that they were, nevertheless, close friends, meeting occasionally at Trefecca with the saintly John Fletcher of Madeley. This work was supported by the Countess of Huntingdon.

Servants of the Kingdom are both greatly used and challenged when they reach senior positions in the world. Their witness in the kingdom acts like salt unless the salt loses its savour. Joseph and Daniel were good Old Testament examples. The General in command of the British army at the present time is an unashamed Christian and respected for it. Citizens are to respect those in authority over them. They are subject to those secular powers that are ordained of God. The weapons of their warfare are not carnal but are mighty, to the pulling down of strongholds such as when it becomes necessary for them to obey God rather than man.

Consider how Jesus came to regard people he met up with. Even though he was God and all knowing there was nothing like hands-on experiences in the flesh. After all he suffered when he was tempted, although without sin, and he took upon himself the very nature of a servant. He was surprised by both the centurion and the Syrophenician woman. Companionship meant a lot to him. As well as being reared in a godly family he enjoyed the hospitality of a happy household in Bethany. He dined with his immediate followers and told them he would drink the fruit of the vine with them anew in his Father's kingdom. He must have told his Father about these things when he drew aside for prayer on his own. His humility was a kingdom model.

An educated young lady applied to serve on a missionary ship only to be told that the only vacancy was for someone to clean the toilets. She accepted. It was not long before she became a useful member of an outreach team. King David said he would rather be a doorkeeper in the house of the Lord than to dwell with the wicked.

This leads to different aspects of service. It concerns those who have portrayed kingdom service characteristics, contributed to God's will being done but not being professing Christians. This does not, of course, include secret disciples such as Gamaliel and Joseph of Arithamea. One hopes that Simon of Cyrene became a believer. Mahatma Ghandi became an admirer of Jesus and followed his teaching bravely in a number of ways. He once said that if he had ever met a Christian who was like Jesus he would have become one. This was an indictment of some but it is also a pity that he did not meet people like Amy Carmichael and William Carey. Gandhi's policy of non-violence and forgiveness of non-Hindus, and practice of fasting, conform to the characteristics of Jesus' nature. It is no surprise that he was assassinated by a Hindu fanatic for his tolerance of Muslims. Nelson Mandela is not a confessed Christian but obviously realized the accuracy of kingdom teaching during his long and unjust detention. He changed from being a reluctant sympathizer of terrorism to a master statesman whose policies of reconciliation and forgiveness averted a bloodbath in South Africa. Martin Luther King was a legend who brought the Negro people out of a system that made them inferior. He was a charismatic preacher who drew parallels with the Old Testament. Yet his private life was inconsistent with his profession.

There have also been men in Orthodoxy who have contributed to the well-being of many. Pope Leo XIII issued the encyclical of *Rerum Novarem,* and Pius XI *Quadragesimo Anno,* in which they outlined principles of social justice many decades before William Beveridge (British Social Reformer) thought of them. It is irresponsible to label every Pope an anti-Christ. It is also entirely unjustifiable. The Borgia Popes and the perpetrators of the Spanish Inquisition undoubtedly manifested the spirit of anti-Christ but these traits were not confined to Orthodoxy. A Pope is first of all a man reared from boyhood within a religious system. He remains loyal to that system and only God has the right to judge him as a person. However, one must never compromise the true evangelical gospel of the kingdom.

In missionary outreaches it is better to meet men and women as individuals rather than as Muslims, Hindus, or total pagans. It is at a personal level that uninhibited attention can be given to the numinous that is within most of us. That is not to infer that there is no value in servants engaging in cross-cultural relations and learning about the good that resides in the cultures and customs of other nations and faiths. At Antioch Paul did not launch into an attack on Hellenistic deities. Wisely, like a true servant of the kingdom, he picked on their unknown god and revealed him to the Greeks as The Lord Jesus.

All people are human beings, and that was the view of President Kaunda of Zambia in 1986. In 1962 an expatriate civil servant was about to leave Nigeria after six years. He was recognised as someone who had a patently kind and sympathetic attitude towards everybody. The Nigerians instinctively recognised these qualities in some of their former colonial overseers. In this case a sumptuous send off supper was arranged. Prior to this a broadcast on radio had made it clear the guest was a Christian. Present were cabinet ministers, senior Nigerian public servants, and the Master of Ceremonies was the Ambassador from the Ivory Coast. There were also millionaire Italian building contractors. Muslims would have been a majority of those present. The host said, "Our brother is leaving and we are unlikely to see him again. Let us stand up and say The Lord's Prayer." It was a solemn and unexpected moment. At the conclusion the whole assembly formed a large convoy to Ikeja Airport where a midnight plane was to come in from Accra and go on to Rome. Normally only passengers are allowed through the inner barrier to the departure gates but officials were unable to deny access to anybody because the first following individual was the head of customs and the next the head of police! As the expatriate walked out on to the tarmac he heard swelling voices singing, "Rock of Ages!" This incident is a confirmation that it is the practical outward manifestation of the fruit of the Spirit that affects reconciliation and mutual respect. Those who subscribe to this are not very far from the kingdom.

9 | The Kingdom and the Word

"O send Thy Spirit, Lord
Now unto me,
That He may touch my eyes,
And make me see:
Show me the truth concealed
Within Thy Word,
And in Thy Book revealed
I see Thee Lord."
               Mary A. Lathbury

As each chapter is read a common thread becomes apparent, namely the frequent references to the word of God. This word is so vital and inspiring that its importance to the kingdom is now dealt with separately in its own right. To start with, God's word is creative. In the beginning of creation God said, "Let there be light. And there was light." Even to human reasoning a belief that an ever existing divine being created matter by a verbal command is no less feasible than a secularist view that there was a spontaneous explosion from nothing. Those with faith do not doubt that God's word is quick and powerful, and sharper than any two-edged sword. It is a living word empowered by the Holy Spirit. Without illumination by the spirit the bible can be just like any other book. The letter kills but the spirit gives life.

When Dr Hudson Taylor was serving with the China Inland Mission he found that educated Chinese were much more interested in medical books than the bible. One highly respected Confucian leader sought after copies of western medical works. He was promised these on condition that he also accepted a copy of the bible; that he would read it; and also that he would pray for enlightenment before he did so. He obeyed and in consequence found the true God and Jesus whom he had sent. It left him with a worry. This was not the severe persecution he was likely to encounter from his compatriots but the fact that he loved his wife dearly and feared alienating her. He hesitated to tell about his experience. When he finally faced up to it there was a surprise. She said she had known for years that a true God existed. When she was a young girl during The Boxer Rising her village had been raided by insurgents who destroyed all the idols and started to slaughter all the inhabitants. Knowing that the idols were ineffective she called upon an unknown God for help and then hid herself in a cupboard. She was the only survivor. She said she recognised this God as being the same one as her husband had found. This man went on to found about 900 churches.

Without the bible there would have been scant knowledge of the kingdom or for that matter about the Gospel either. It is one thing to have a head knowledge of its contents and quite another seeing it as being a means of finding and receiving a resurrected living saviour. There are numerous testimonies of those who have called on God and been answered in spirit. These include those in acute danger, those who had come to the end of themselves in prison, and others who had simply responded to the exhortation: "They who call upon the Lord will be saved." Psalm 130 has motivated countless people since it was written. "Out of the depths have I cried to you O Lord. Lord hear my voice. Let Thine ears be attentive to the voice of my supplication . . . with you there is merciful forgiveness . . . my soul has waited for you like those who wait for the morning." Jesus said that except a person is born again of the spirit they cannot see or enter the kingdom. Neither can they have any real spiritual appetite for the word. That is why the word is rightfully described as the sword of the spirit. It penetrates to the dividing of the bone and the marrow.

The word does not fully explain the proposition of one God in three persons as it is a mystery. However it does make it abundantly clear how their three separate offices are performed. The Father is the authority or originator. Jesus is described as the logos or living word of God and in John's gospel in particular as the word made flesh who dwelt amongst us. It reveals the Holy Spirit as a person and states that Jesus described him as "He". Nevertheless, the Spirit is omnipresent. In this way the Spirit conveys new life from Jesus as the source to those who abide in Jesus as the vine. Thus the word makes plain to us the dynamics of the kingdom. In itself it is a living word that will not pass away even when heaven and earth cease to be.

People need to have this word explained to them and for this purpose God has set teachers in the church. These have to be faithful not only in teaching but also by offsetting the activities of false teachers.

An Ethiopian eunuch in the service of Queen Candice was reading the 53rd chapter of the book of Isaiah in his chariot and could not understand it. Philip was transported by God alongside. The prophecy was made 700 years before Christ's first coming. When Philip asked him whether he understood what he was reading the eunuch said this was not possible without help. The passage read, "He was led like a sheep to the slaughter, and as a lamb is dumb before his shearers is silent, so did he not open his mouth. In his humiliation he was deprived of justice. Who can speak of his descendants? For his life was taken from the earth." Philip was able to say that this was about Jesus who had been crucified recently and that it went on to reveal how he had borne our sins. The Eunuch gladly accepted this and asked Philip to baptize him. This is a very good example of how citizens must be able to open the precious word to others. The word asks, "How can they call on the one they have not believed in? And how can they believe in one of whom they have not heard? And how can they hear without someone preaching to them?"

Citizens must not confine their Christianity to Sundays. They must study to show themselves as workmen approved of God and must always be ready to give an account of the reason for the hope that is within them.

Many are so convinced of the effectiveness of the bible to speak for itself that there are many organizations that exist just to distribute it far and wide in different countries and languages. These include those who continue to provide translations in remote tribal languages. There are many testimonies from those whose lives have been changed simple by reading alone.

There are misunderstandings about the word of God. It is asked how so much store can be placed on one book when there are so many works by renowned philosophers and poets? There are books by Plato, Aristotle and Sophocles, and later ones by Kant and Freud. Surely, it is not possible to rely, some say, on what are translations from other languages (i.e. Hebrew, Greek, Latin), and from past and foreign cultures. It is, therefore useful to be reminded about the composition of the bible. It has sixty-six separate books written over a period of one and a half millennia. It is still the most widely distributed of all books although it was completed two millennia ago. One organisation distributes one million bibles or testaments every week. The writers had varying backgrounds and disciplines, and would not have been aware that when the canon was complete it would present a consistent doctrine. The New Testament has three synoptic gospels: Matthew, Mark, and Luke which are largely a Life of Christ. In the fourth gospel (John's) there is a doctrine of the Logos, and a spiritual insight into the nature and mind of Christ. Besides the four gospels there are fourteen epistles by Paul (including Hebrews) and eight books that are the works of four other authors. Revelation was the earliest of these and merits it own chapter. The other seven books were addressed universally – not to a city or town.

Some modernist faculties teach that Paul invented Christianity. This is palpably untrue. Paul went more deeply into doctrines than others but there is no inconsistency between his writings and those of, say, John or the sayings of Jesus. His depth has given scope for others to make more of his teaching than is warranted, particularly as to "predestination," and "election", and the future of historic Israel. Paul's writings can be recognised as consistent and God inspired. However, there are some , including liberals in the broad church, who spend a lot of time criticising these writings together with the rest of scripture.

There are two other things about the word worth noting. It is a guide against false prophecy and teaching. The spirit and the word must agree. Then there is a promise of the immortality of the word. "Heaven and earth will pass away but my words will not pass away."

As the influence of secularism has increased so has antipathy toward the word of God and towards evangelical or bible based Christianity. Under the guise of political correctness there are objections on the part of corporations to the display of Christian symbols and sometimes the placing of bibles in hotels, schools and hospitals. There are more relaxed attitudes towards the requirements of other faiths which shows that although there is a welcome liberalisation at work in general there is an unwelcome discrimination in some quarters against Christians. Strangely, this does not seem to apply to the prison service perhaps because it is seen as a last resort in tough cases. The practice certainly bears fruit. Other faiths dissent from a public view that they dislike Christian witness.

In order to offset some unfavourable comments about the traditions of orthodoxy it is good to mention welcome trends. Cardinal Newman's comment on the medieval era cannot be applied to the present day. It is not necessary to agree with others' doctrinal opinions in order to say things that are favourable. With this reservation it can be denied that the Catholic church has not changed at all. The change is an openness towards the scriptures and making these available to adherents. This is a welcome reversal of secular attitudes. It would have been unthinkable in the past for a non-Catholic organisation to be allowed to present New International versions of scriptures to Catholic schoolchildren. The following is the gist of what was said to boys at a Catholic secondary school morning assembly. "This version (NIV) was translated by over 100 scholars from Hebrew, Greek and Aramaic texts. Your head teacher and your teachers consider it is good to read. But how does it compare with Catholic versions? Your Douay, Knox and Jerusalem Bibles based on Latin Vulgate diligently compared with original Hebrew and Greek texts are equally faithful and accurate. There is no marked difference in the sense at all except that these contain a few extra Old Testament books in the canon. When visiting some state schools we have to explain the value and importance of the scriptures with carefully chosen words. We can be much bolder and direct in recommending this book to you all.

The Second Vatican Council said of the scriptures what Catholics should do, namely: put writing as under the inspiration of the Holy Spirit; read them frequently because they tell about Jesus. To be ignorant of what is in them is to be ignorant about Jesus (Jerome)."

Leo XIII said the bible "was a heaven sent treasure", and Pius XII said it was "inspired by The Holy Spirit and works to the supernatural salvation of souls. Peter said: Lord to whom shall we go – Thou hast the words of eternal life."

A senior member of Opus Dei recently went to speak at a school and he spoke of how Jesus was his Saviour. He was very well received and said he thought the pupils were surprised to find that he did not have horns! So through the anointed word the kingdom is advancing in many quarters even if it (the word) is being banned in others. It is all part of the ongoing battle between the light of the gospel and the darkness of the world.

> "God's Word, for all their craft and force
> One moment will not linger,
> But, spite of hell, shall have its course;
> 'Tis written by His finger.
> And though they take our life,
> Goods, honour, children, wife,
> Yet is their profit small:
> These things shall vanish all;
> The city of God remaineth."
> *Martin Luther, translated by Thomas Carlyle*

10 The Holy Spirit and the Kingdom of God

"Eternal spirit of the chainless mind!
Brightest in dungeons, Liberty! thou art,
For there thy habitation is the heart."
Sonnet on Chillon, Byron

Much has been covered already about the work of the Holy Spirit in bringing people into the kingdom by new birth mainly in Chapter 5. It remains now to deal with that part of his office that is concerned with activating God's rule on earth. This work in the setting of the three persons of the Godhead can be seen as authority coming from the Father, redemption and eternal life as from the Son, and the actual carrying out of the divine counsels by the Holy Spirit. When Jesus was on earth he was necessarily confined to one location at a time. John's Gospel states that he was the word made flesh who dwelt amongst us (or tabernacled, or was "God in a tent"). He was the sinless Son and the first born of a new creation. However, before he started his public ministry he went to the river Jordan for water baptism by John the Baptist. It was then that the Holy Spirit descended upon him for empowerment. He was immediately driven into the wilderness and withstood three temptations by the devil by using the word of God, which is the sword of the spirit. As he fasted for 40 days the devil tried to get Jesus to perform a miracle by turning stones into bread when he was hungry. Jesus said that man should not live by bread alone but by every word that proceeded out of the mouth of God. He then enticed Jesus to throw himself from a pinnacle so that his father would rescue him and he would be proved to be the Son of God; Jesus replied that he should not tempt the Lord his God. The devil then offered authority over the kingdoms of the world if he would worship him and it must be assumed he was able to do this. Jesus then said it was written that only the Lord his God was to be worshipped and Him only should be served. An empowered Jesus withstood these cunning suggestions and this is a salutary example of how empowerment of the divine nature in citizens enables them to be spiritual warriors against the wiles of the devil.

The kingdom was at hand or imminent when Jesus was physically present and his objective was, with the anointing of the spirit, to preach the good news to the poor, to proclaim freedom for the prisoners, and recovery of sight to the blind, and to release the oppressed. He began by stating that the Spirit of the Lord was upon him. This was a direct quotation from a prophecy of Isaiah read by Jesus in the temple. It is noteworthy that Jesus cut off at "to declare the day of the Lord". By leaving out "The day of vengeance of our God" he showed that his first coming was not to condemn the world but that the world through him might be saved.

Jesus explained that it was expedient that he should go away out of the world so that the Holy Spirit would come as his Father had promised him. The Holy Spirit would be omnipresent (he would be everywhere) and would not seek to glorify himself. He would glorify Christ. He would take of the things of Christ and bring them to citizens. Having already brought them to new birth he would enable them to remain in Christ, the vine, by conveying fresh supplies of grace. Having the very nature of Christ they could be empowered for the service of the kingdom as Christ was at Jordan. Only Jesus, however, had the spirit without measure. He gave authority to his immediate apostles and disciples to preach the kingdom and to cast out demons. From Pentecost onwards citizens of the kingdom could be endued with the power of the spirit. They could receive the gift of the Holy Spirit. John the Baptist explained this by saying that it was Jesus who was the baptiser with the Holy Spirit and with fire. Peter explained on the day of Pentecost that it was a fulfilment of a prophecy of Joel that in the last days God would pour out his spirit on all flesh and that sons and daughters would prophecy. The gift was for their children and children's children even to as many as the Lord our God would call. It should be noted that Peter put the time as the beginning of the last days and said it was for those who were called. So Jesus was the baptiser and the Holy Spirit the conveying means.

There are various ministry and spiritual gifts that are imparted and set by God in the church for the service of the kingdom. It is useful to reflect on the difference between fruit and gifts. Gifts are for the equipment of citizens for use in the present spiritual invisible earthly kingdom. They are not for eternity. On the contrary fruit is from the spirit of life in Christ. It stems from the outworking of the new nature and is eternal. It is also sanctifying. When citizens meet Christ they will no longer have sinful natures. When they see him face to face they will be like him. Beholding him they will be transformed into his likeness from glory to glory as by the Lord the Spirit. Now it will be realised that the same spirit gives power for witness and conveys the Christ life. The mystery hidden from the ages is Christ in us the hope of glory. In addition he is also a comforter and an aid in intercession. He is not a mediator or an advocate. In addition to strengthening citizens the Holy Spirit also combats the powers of darkness in direct ways such as by convicting the world of sin, righteousness, and judgment. When he brings revivals about, this is particularly evident. Paul described the power as equivalent to the mighty strength exercised in Christ when he was raised from the dead.

The spirit is given so that all can profit. This can be a disconcerting matter for the organisers and controllers of formal religion. It so frightens the devil that he concentrates his attacks wherever there is liberty in Christ, regrettably to great effect. He is certainly not afraid of bells and smells. Revivals occur from time to time. Small ones are much more frequent than some think. Amongst notable ones Ballymena (Northern Ireland) in the nineteenth century saw pupils in two adjacent schools being convicted spontaneously and corporately. Simultaneously adults were struck down in the street as they crossed a certain point on the pavement. These events resulted in repentance and changed lives on the part of many. A rather different revival spread across the United States of America and became known as "The Great Awakening". This began when a shortsighted unprepossessing man read out a written sermon. At words like, "As for me my feet had almost slipped; I had nearly lost my foothold," about three hundred people cried out to God for mercy. There was a revival in Wales in 1904. Thousands came into the kingdom. New chapels were built and filled all over the country. Many coalminers were converted. Because these men stopped swearing the pit ponies no longer responded to their commands! It was said that heavenly singing could be heard coming from the sky. The fire of this revival died out as quickly as it came but its fruit in conversions and the establishment of chapels did not. A leader went ahead of the Holy Spirit and claimed that there would be a materialisation on a mountain top of the Holy Spirit. Crowds gathered but nothing happened. The Holy Spirit withdrew the fire.

More will be written later about the spoiling affect of Satan amongst citizens. King David prayed that God would not take his spirit away from him and this is a good example of the difference between salvation which is not lost and the power of the spirit that can be withdrawn when grieved, just as the spirit left the Jewish Temple.

There was a more recent revival in Cambuslang in the Scottish Highlands. This began in a small way in Lowestoft and was carried by fisher people further north. Prayer and fasting are thought to be instrumental in bringing about revivals and this gels with the Holy Spirit being a spirit of supplication. When citizens do not know what to pray for as they ought, the spirit intercedes with unutterable groaning. Through praying in the spirit one is able to discover personal shortcomings and incorrect priorities.

Our subject does not call here for a detailed treatise on gifts. A whole book would be a minimum requirement for such an undertaking. Their place and use in the work of the kingdom and why and how there is opposition to their operation today is very relevant however. As to ministry gifts in the church there are those who study with a view to entering into full-time or part-time service in the church. Inevitably other adherents, the sheep, look up to them. It is biblical that God sets in the church apostles, prophets, evangelists, and teachers. There is no mention of priests. It can be said that, generally, the sheep are not too conversant with theology. Their understanding is shaped on what they are taught. As people have ordinary lives to live this is a good arrangement provided that overseers are genuinely godly and see their roles as those laid down. God's people are to be prepared for service so that the body of Christ may be built up until a unity of the faith is reached in a knowledge of the Son of God and become mature, attaining to the whole measure of the fullness of Christ. This is a most important aspect that is often either neglected or soft-pedalled. It does mean that in many denominations people are taught things that are not biblical and it has led largely to the organized visible churches containing mixtures of citizens and strangers, the latter not realising what they are not. There is sometimes a lack of involving the flock in the work.

It is clear that in apostolic times the operation of both ministry and spiritual gifts flourished. It is contended that potentially these could and do sometimes operate today subject to constraints. These range from those imposed by the Lord for various reasons; the opposition of legalists who use a number of arguments, and lack of sanctification and faith in individuals. However, ideally the orderly operation of gifts is an asset to the kingdom because this is why they are set in the church or imparted. God is not prepared to show signs to any wicked and adulterous generation because he will not allow his obvious miracles and supernatural manifestations to by-pass a need for repentance and acceptance of the gospel of the cross. Nor can God give credence to an unclean vessel by allowing its ministry to be confirmed with signs following. Legalists oppose the operation of gifts today because intellectually the supernatural rankles with them. Also a number of doctors of the church were, and are, understandably, wary of counterfeit and hysteria. Their mistake is in finding intellectual reasons for the cessation of gifts, or of those they dislike. They limit inspirational gifts on flimsy grounds such as that the canon of scripture is complete or that they were intended for the apostles and a few others. There is no sound scriptural backing for these arguments. More erudite and spiritually minded expositors stick to scripture and are not prepared to support unconvincing arguments.

To the consternation of extreme charismatic sects the great English theologian John Owen wrote that in his day (seventeenth century) spiritual gifts had largely ceased – adding "that is not to say that God will not revive them in the future, or that in the past he has actually done so." Having brilliantly discerned the true facts and failing to find any scriptural justification for cessation he then let his spiritual guard down and offered opinions as to why some gifts were redundant. He excluded some gifts and put others in strange wrappings. He has done us a favour by demonstrating the folly of going by experience at a point of history and not by the bible which he was otherwise a master of. In other words he put two and two together and made five! As an example he said there was no longer any need for the ministry gift of an evangelist. This conclusion was arrived at in Cromwell's times when England was under tight Puritan control and there was no British Empire or missionaries. With the benefit of hindsight it can be seen such a view was asinine. What does this make Billy Graham? He handled the gift of prophecy in an equally strange manner. Philip had four daughters who prophesied: "But they were not prophetesses" – which is a correct statement. However he went on to state that, "from time to time they had revelations, for prophecy is nothing but to declare hidden and secret things". But surely that would make them prophetesses? His mistake was to lean on his own understanding. The bible makes it clear that the New Testament spiritual gift of prophecy is purely for edifying the church. Citizens are to prophesy according to the measure of their faith. All can do this providing that they abide with the rules of order. This gift was confused by Owen with the word of knowledge that would have been used by Agabus. The gift of prophecy was not to be despised. It was equivalent to speaking in tongues with an interpretation. Philip's daughters edified the church. If this is accepted then a biblical admonition that people should prophecy according to the measure of their faith makes much more sense. Then there is Owen's opinion that it is now confined to teaching and preaching. It is true that a preacher can be aided by this gift and it is not difficult by means of ordinary discernment to judge when this is and is not the case. Now this case has not been included to disparage the ability of an outstanding theologian. For it is not the purpose of this book. It is sought rather to show how natural reasoning is used to inhibit the work of the Holy Spirit in the Kingdom of God and bring out how yet again an aspect of the kingdom is being hidden.

A modern preacher hit the nail on the head. He was asked to take a service in place of an absent minister in a church and town that were unfamiliar to him. On the subject of the Holy Spirit he said that on the day of Pentecost Peter explained the strange phenomena as, "This is that which was spoken of by the Prophet Joel." He went on to say that balanced Christians had every right to say of some of what was happening, 'This is not that'. whereupon the congregation seemed pleased. He continued by warning that great care should be taken to avoid saying "This is not that" about the true operation of the Holy Spirit who would be with God's people until the end of the world. This illustrates the crux of a contemporary problem. Unknown to the preacher there had been a rift in the church. The half of the congregation who were charismatic had left. Even in many "charismatic" churches a control is exercised to prevent uncensored participation of the flock in worship. In those that are more open there is danger of things getting out of hand. A solution is to have godly and experienced people in the oversight with discernment. Until it is recognised that the devil works overtime to prevent real liberty in the spirit when there is a coming together there is unlikely to be a weighing up of whether it is right to quench the spirit in the interests of what is perceived to be sobriety.

The bible gives plenty of warnings about false prophets and deceivers and these abound at the present day. It is illogical that the Father promised his Son to send the Holy Spirit to be with the church to the end of the world and leave in his word detailed teaching about how the spirit will operate, and then almost at once irrevocably limit his power. As has been explained power has been withdrawn in some respects over periods on account of human failure. There are recurring problems with the abuse of spiritual gifts and these disruptions are largely engineered by the devil. In the long run he will not succeed in thwarting the advance of the kingdom as God wills. God's servants are weak in themselves. They have this treasure in earthen vessels in order that the excellency of the power is of God and not of ourselves. He gives to those vessels who are humbly obedient mouths that cannot be withstood. They do not need to use enticing words of human wisdom. Ministry gifts of the spirit are of paramount importance. As to spiritual gifts, it must be understood that the new testament gift of prophecy (i.e. edifying the church) is not revelatory, that genuine spiritual gifts are not currently common place, but they have not ceased altogether in the divine calendar.

Apostles plant churches today just as Barnabus did and Barnabus was an apostle although not one of the twelve, and neither was Paul. Evangelists, pastors and teachers either carry the gospel to strangers or build others up to do the same. It is a consolation that when the enemy comes in like a flood the spirit of the Lord will raise up a standard. The power of the Holy Spirit was party to establishing the enclave of the Kingdom of God on earth. Prayer must be offered continually that God's servants will be mighty through him to the pulling down of strongholds, casting down imaginations and everything that exalts itself against the knowledge of God, and bring into captivity every thought to the obedience of Christ.

11 | Evil and the Kingdom of God

"A safe stronghold our God is still,
A trusty shield and weapon;
He'll help us from all the ill
That has now o'er taken.
The ancient prince of hell
Has risen with purpose fell;
Strong mail of craft and pow'r
He weareth in this hour;
On earth is not his fellow."
Martin Luther

If anyone has a vested interest in hiding the Kingdom of God as a present dynamic reality it is the Devil. Just as he brought about mankind's fall by his temptation of Eve, he continues to deceive Christ's church. By the turn of the second century AD he had already made grievous inroads, and more about this can be read in chapter 16. In the second of his letters to the newly established Corinthian church the apostle Paul wrote, "I am afraid that just as Eve was deceived by the serpent's cunning, your minds may somehow be led astray from your sincere and pure devotion to Christ. For if someone comes and preaches to you a Jesus other than the Jesus we preached, or if you receive a different spirit from the one you received, or a different gospel from the one you accepted, you put up with it easy enough." He went on to describe the deceivers as "super-apostles". We encounter these in our churches today and you can read more about the problem in chapter 19.

The Devil is never averse to hiding himself while remaining in wait.

In this way he can cause people to be complacent. Those who do not believe in God find it comfortable not to believe in the Devil either. As they grow older all there is to live on is past memories. Those who have been dishonest and have bad consciences wish the clock could be put back. Lord Byron was one such and his despair is reflected in his poetry: "Oh could I feel as I have felt, – or be what I have been, or weep as I could once have wept o'er many a vanish'd scene; As springs in deserts found are sweet, all brackish though they be, So midst the wither'd wastes of life, those tears would flow to me." At least his conscience was not seared with a red hot iron as is the case of diabolically evil people. This poem was preached on at a conference and as a result a man went to his employers and confessed to embezzlement. A brand snatched from the burning! He said he had been showed that he had been shovelling coals for the devil for years. The devil is, at this time, the acknowledged Prince of this world, and, unlike God's kingdom, this is visible. Whenever he or his evil associates do come out of the woodwork there can be evil manifestations.

It can be gathered from scripture and from actual experiences that divine and satanic powers impact upon people each in a different manner. The power that comes from above meets firstly the spirit, then the soul, and then the body. That is the order. Evil power is earthly, sensual and devilish and meets body (flesh), soul and spirit in that order. Since these powers are quite the reverse of each other it can be seen which will cause hysteria, emotionalism, and sensual behaviour. God is not the author of confusion. Satanic power so affects the flesh that it can cause "goose pimples" and extreme shivering even in an equatorial climate. Satan is not a black figure with a three-pronged pole. He comes as an angel of light that is luminescent. A brush with satanic power has been known to convince an atheist of the existence of the supernatural and since this was of a counterfeit nature also a realisation of true divine power. Evil is most prevalent in totally godless environments. Some countries are pagan and plumb the depths of spiritual wickedness. People are oppressed, maimed, and young children are forced to become mercenaries and murderers. Genocide has become a new word in dictionaries. Bearing in mind that Satan is bent on man's destruction it is not surprising that human sacrifice, and cannibalism, still feature in primitive societies although it is supposed to be illegal. In civilised countries the attack is upon those who are drug addicts or trapped in occult practices. In cases of suicidal attacks on others such as with Kamikasi pilots or Arab insurgents the losers are both the perpetrators and the intended victims. Satan is the winner.

In the early hours of a Sunday morning two young bank workers sat in a car listening to heavy metal music played backwards. A hose led from the exhaust to the inside of the vehicle. The engine ran and brought death to both of them. Just across the pavement stood a chapel. That evening as it got dark one set of parents came to place flowers. Lights were on in the chapel and they were drawn to go inside and sit at the back. The preacher was aware of the tragedy but not of the parents' presence. He prayed feelingly for them and preached a customary gospel message. The parents were touched by the service and asked to be counselled. The parents were shown scriptures of comfort and subsequently joined the chapel. This was a delicate task almost spoiled by an erstwhile interloper who said very loudly, "Well he is in hell then!" The preacher was asked to conduct the funeral. The young man had been popular and played American football. The crematorium was packed out with young friends on the day. The family asked that attention should be drawn to the dangers of the occult in which the two young men had dabbled. There had been an abnormal number of suicides by young people in the town. Reports from America revealed suicides and attempted suicides by heavy metal fans. Amongst the scriptures quoted was, "So that by His death He might destroy him who holds the power of death that is the devil – and free those who all their lives were held in slavery by the fear of death. For this reason he had to be made like us in order to be a merciful and faithful high priest in the service of God and that he might make atonement for sins of the people – because he himself suffered when he was tempted he is able to help them who are being tempted." Having spoken about the young man's life the preacher broke off from the theme of the funeral and said, "From what I have said there seems no sense in the ending of this life and it would be pastorally wrong and intrusive to say anything more about it. I must say that Christianity that has a form of godliness but denies the power thereof is failing young people today. Because society is not meeting their deeper needs they hunger and thirst after satisfaction that cannot be found in a materialist world. They are let down by those who scorn the deep and immensely satisfying experience of finding the saving and fulfilling power of a risen saviour in the person of Jesus. This is being preached in live churches. Please seek one out if you have been affected by what I am saying. I was in a self-righteous trap of formal religion until my mid thirties and it was only then that I found through the scriptures that true Christianity was repentance to God and faith in Christ leading to a new life and hope in the promises of the living word of God. I sought God because things were not going right. Thank God I was not confronted with present-day obsessions such as psycho-spiritual frontiers of human experience and so-called theoretical physics.

Jesus said that Satan was a deceiver from the beginning. May God rebuke all evil and deliver us from it. I urge all young people not to be deceived by what is fed to them in the world but to seek after the true and living God while he may be found. You are going to miss this young man when he is no longer at the desk or on the field of play or wherever you used to see him. Let me quote from the poet Thomas Gray:

> "No farther seek his merits to disclose,
> Or draw his frailties from their dread abode,
> (There they both alike in trembling hope repose,)
> The bosom of the Father and his God."

So this was a real encounter with the Satanic, as opposed to a welter of pretence, imagination, fear induced by others, and counterfeit practices. Pretence to Satanic power can, nevertheless be evil in itself. Witchdoctors in Africa claim they can raise the dead from the grave. This strikes terror into the populace because unlike the case of Lazarus the claim is to raise them in a semi-decomposed state. This is a very gruesome business A respected and responsible Nigerian public sector executive testified that one dark night he was followed by one of these apparitions and fled in terror. Nobody doubted that his experience was genuine. He saw, when he looked back, something like a walking corpse covered in flies. Some years afterwards it was established that certain witchdoctors could take a person captive and by administering drugs and potions and concoctions produce a human zombie. What the executive saw was not a raised corpse but a dehumanised being. In the 1930s a family living in Harrow became involved with a sinister wealthy man from London who was a medium. Séances were held in the family home and the youngest son was possessed. He was a slight figure, an arts graduate, and a professional organist. It took a 17 stone rugby player and a policeman to restrain him and get him in a straightjacket. He said alternately that he was Jesus Christ and the devil. He was never able to follow his occupation on eventual discharge from an asylum and worked as a night telephone operator. The head of the family said that there was some kind of materialisation. The following account throws more light on this sad case. *None Other Gods* was a best selling book of the 1930s. It was later adapted for the silver screen under the title of *The Necromancers*. A young man lost his fiancée as the result of a tragedy and was persuaded by Spiritists that she could be contacted in a séance. What followed was an apparent materialisation that he ran to. He was then possessed by a demon. Now this book was written by Hugh Benson, a Roman Catholic priest. The then Provincial of the

Dominican order said Father Benson had confirmed that this was not fiction and had introduced a character in the book to him at Euston station.

God is not the author of magic, pointless manifestations, divination, sacrifice (except the one true sacrifice), levitation, poltergeists, ouija boards, and clairvoyance. These are all part of the devil's spiritual wickedness. These are not new things. After all Satan has lived a long time. So there was the witch of Endor who raised the spirit of Samuel, the offering of strange fire by the sons of Aaron who perished, and the seven sons of Sceva who had all their clothes torn off by evil spirits when they attempted to cast out without authority.

There is a major religion that uses a calendar based on astrology as opposed to astronomy. Leading astronomers know more than most people that astrology is worthless. One Guru who led a breakaway cult with a following in America altered the official calendar so that it gave the date of his appointment more "authenticity". But horoscopes are going to require a whole chapter and this follows. It has been attempted here to give examples of evil powers that opposed the advance of the Kingdom on this earth. The bible tells us of principalities and powers and wickedness in high places. These are largely secret. One day everything that is done in secret will be revealed at the judgment.

12 | Horoscopes

"Various trees corresponded to cardinal points, and the old gods and heroes took their places gradually in a symbolic fabric that had for its centre the four talismans of the Tuatha de Danaan, the sword, the stone, the spear and the cauldron which related themselves in my mind with the suits of the Tarot. George Pollexfen, though already an old man, shared my plans, and his slow and difficult clairvoyance added certain symbols. . . . The forms became very continuous in my thoughts, and when AE came to stay at Coole he asked who was the white jester he had seen about the corridors. It was a form I associated with the God Aengus." *W. B. Yeats, Autobiography*

Folly in Reading Horoscopes

An addicted populace It has been estimated that over two-thirds of the adult population of the British Isles read horoscopes and that about half of this number actually believe in them. The other half do so with tongue in cheek, but all are prey to what has now become a worldwide multi million pounds industry. Even the more reputable parts of the media are now including astrological features. Horoscopes are based on astrology: that is a belief that by study of the stars it is possible to predict future events and to discover the days and times which are "lucky" for individuals. Some people see this as exciting and as a welcome distraction from monotony. It is sad that they do not, instead, seek after a true knowledge of the living God. Their choice leads them into making wrong decisions about the course of their lives and putting superstition above rational thought.

Need for scrutiny It is up to citizens to know about horoscope reading so that they can tell people just how harmful this practice really is. It is quite impossible that the position of planets and stars billions of miles away from the earth can have anything to do with future events or, for that matter, with the temperaments of individuals born when planets and stars are in a particular position in the heavens. Any unbiased study of the results of horoscope predictions will show that most of them are wrong. The few that actually come to pass can be seen to be the outcome of pure chance. Great publicity is given to the so called "successes" whilst the far greater number of failures is conveniently overlooked. The process is like the shady world of race horse tipsters whose claims to have predicted the winners ignore the far greater number of also rans. It is just a statistical fact that large samples of predictions throw up "successes" that are entirely fortuitous. Both race tipsters and casters of horoscopes would have no need to solicit for money if their claims were genuine. They would be able to make money for themselves. Before the last world war virtually all horoscopes predicted that there would be peace. This was because modern astrologers told people what they wanted to hear in order to keep them happy and paying up. How wrong they were! Ancient Babylonian astrologers, on the other hand, tended to predict woes and disasters and were woefully wrong.

A flawed black art The rationale upon which horoscopes are based is a flawed one. The exact science of astronomy has exposed the falsity. Pagan astrologers thought long ago that there were seven planets that revolved around the earth. These were named: moon, mercury, mars, venus, the sun, jupiter, and saturn. They were named after "Gods" and "Godesses" of mythology. This in itself is against the first commandment of God – "You shall have no other Gods before me" (Deuteronomy 5:7). Astronomers have discovered during the past 250 years that all planets revolve around the sun, and not the earth as pagan astrologers supposed. The earth is a revolving planet as well, and there are three more planets that were hitherto undiscovered: Uranus, Neptune, and Pluto. So what does this do to upset astrologers' concepts? It was known that the planets (or moving stars) always kept within a narrow band, called the zodiac, of fixed stars. These fixed stars were divided into twelve "houses" or "constellations." by astrologers and the houses were given names. The names were The Ram, The Bull, The Heavenly Twins, The Crab, The Lion, The Virgin, The Scales, The Scorpion, The Archer, The Goat, The Old Man with the Water Pot, and The Fish with the Glittering Tails. It seems ridiculous to suppose that planets that are passing through these concocted houses when a child is born can influence its character and it future. Such superstition is made worse by claiming that the planets and zodiac signs have powers corresponding to their names. The weird thing about these "houses" is that these only seem to be on the same plane as one another as viewed from the earth. In reality the nearest star is twenty-five billion miles away, the brightest star, Sirius, is twice that distance away, and the nearest nebula is 100,000 times that distance away!

Horoscopes are forbidden by the Bible Horoscopes, as has been said, are based on astrology. Apart from astrologers' disobedience to the first commandment of God, there are other scriptures that make it abundantly plain that their practices are forbidden and that they are closely linked with necromancy (communicating with the dead), divination, sorcery and fortune telling. Here are some of the things that the Old Testament has to say: astrologers and stargazers who make monthly predictions are in error. They cannot save themselves or save others (Isaiah 47: 13–15). There is a warning not to learn the ways of nations or be terrified of the signs in the sky (Jeremiah 1:2). Astrologers were completely ineffective to predict or make revelations in Daniel's day, whereas Daniel, in whom the spirit of God indwelt, could (Daniel 1:20, 2.27, 4.7, 5:7, 8). Sorcery was useless (Isaiah 19:3–11). Sorcery was inferior to God's power (Exodus 7, 11, 22). Necromancy was forbidden (Deuteronomy 8:11, Isaiah 8:19). Astrology was not credible (Job 38:31–33). In the New Testament one encounters first of all the intriguing case of the Magi, or the wise men from the East, who followed the star to Bethlehem. It will be necessary to face up to counter-claims that these Magi were in fact astrologers of the Magian religion who saw the star as a conjunction of Jupiter and Saturn in the year 1 BC. However, this explanation is at odds with the second Chapter of Matthew's Gospel which contains the only account of this event. The obvious interpretation of this text is a supernatural appearance of a meteoric star and not an ordinary star of the created universe. It is highly likely that inter-cultural relations between Jews and Persians during the captivity and the subsequent residence of Jews of the dispersion in Persia (see Acts 2:9) may have made the Magi familiar with Jewish beliefs about the Messiah. It is quite certain that the Magi were given direct revelation from God as well as to interpreting the sign and directing their course of action as indicated in Matthew 2:12. Mr. Pritchard in Smith's *Dictionary of the Bible* remarks "Arrival on the hill and in the village, it became physically impossible for the star (i.e. the conjunction of Jupiter and Saturn, supposed by Kepler and others to be the star) to stand over any house whatever close to them seeing it was now visible far away beyond the hill to the West and far off in the heavens at altitude 57+. As they advance, the star would of necessity recede, and under no circumstances could it be said to 'stand over' any house unless at a distance of miles from where they were. A star if vertical would appear to start over no house or object in the immediate neighbourhood of the observer. This beautiful phantasm of Kepler and Ideler which has so fascinated many writers, vanishes before the more perfect daylight of investigation." Mr. Pritchard's explanation is clumsy. He seems to imply that the star must have been supernatural? There is ample other evidence in the New Testament of Christians' repugnance of astrology and other dark arts. There are the instances of the girl with a spirit of divination which Paul cast out (Acts 16), of Elymas the sorcerer God made blind for a season as a punishment (Acts 13), and of Simon Magus who made the mistake of supposing, because of his occult background, that spiritual gifts could be purchased.

Occult connections There are occult links with horoscope casting particularly in relation to fortune telling. Evil spirits impersonate when attempts are made to contact the dead. King Saul incurred God's anger by going to a witch in order to contact the spirit of the dead prophet Samuel (1 Samuel 28). A main trait of Satan and his wicked servants is to deceive and lie. Jesus said that false shepherds only come to steal, kill, and destroy. But Jesus as the Good Shepherd laid down his life for the sheep. He came that we might have life and have it more abundantly. He came to free us from bondage, fear and superstition: to release the prisoners. This included freedom from the bondage of reading horoscopes. By his death he overcame the works of the Devil and purchased eternal life for those who would believe in him. Because of the connection between horoscopes and the occult Citizens should pray against them and for those who are caught up in such a superstition. It is not only unintelligent people who are caught up in the dark art of astrology. World leaders both past and present have been affected and the affairs of nations misdirected. W. B. Yeats, the distinguished Irish poet, dabbled both in spiritism and in horoscopes. Two sample extracts from his diaries read: "Dec 16. I should have noted that I had and felt a good deal of gloomy anger some troubles. Jan 11: or day before or after will bring some trouble, a quarrel of some kind." He entered astrological symbols as a shorthand for: Mars in opposition to Saturn and Mars conjunct mid-heaven progressed. Numerous examples of such notations are to be found in his diaries and astrological work practices. It is sad that quite brilliant people can be taken in by superstitions of this kind. It was supposed that before Julius Caesar was assassinated that a soothsayer uttered, "beware the Ides of March". Hitler's occult advisors were either bogus or else the agents of Satan driving him to a final suicide. He was an example of a demonic man in his own right. Citizens are told to resist the devil and he will flee from them. On no account are they to rebuke him because he is still a powerful angelic being. Even the Archangel Michael durst not rebuke him but said, "The Lord rebuke you." Citizens should realise that credence of such things stretches into high places. Principalities and powers are often governed by twin evils of greed for wealth and position, and involvement in the occult. In the same way that the kingdom remains hidden to many, so Satan's major adherents meet in secret places that are hidden from the outside world. Not many strangers hate the kingdom of God but those who do hate it hate it heartily and conspire against it with the help of Satan and his demonic forces. Any regime that is unfairly oppressive either by using force or by exercising discrimination either overtly or secretively must belong to the world of darkness. The news of the kingdom must be uncompromising because it is true and liberating. Other religionists and persons who practice different standards of moral conduct are particularly offensive to God when they persecute his people with violence or deprive them of freedom of speech. It follows that if Christians ever portray such traits they are not manifesting the Spirit of Jesus.

13 | Hidden from the Wise

Jesus said, "I thank you Father, Lord of Heaven and earth, because you have hidden these things from the wise and revealed them to little children." *Matthew 11:25*

This statement was not just a piece of inverted snobbery on the part of Jesus. His incisive, divine mind, knew what was in man. In this context Jesus was referring to those who either would not accept the miracles he performed or else attributed them to demonic forces. They would not accept his divinity and found fault with the company he kept. It was they who were the snobs, as they considered themselves superior. On the other hand the poor heard him gladly. There is a spiritual wisdom, such as Solomon's that is God-given. There is nothing wrong with natural wisdom boosted by education and background so long it is applied properly. That is why Paul advised, "be careful how you live – not as unwise but as wise – making the most of every opportunity because the days are evil." Three of the disciples were fishermen – Andrew, Peter and James. Three others were respectively a doctor, a civil servant, and a religious scholar – Luke, Matthew and Paul. They all depended upon the power of the Holy Spirit, something that today's 'wise' men call "enthusiasms". These so-called wise men have an agenda of their own. This will be described briefly without too much comment because readers are expected to exercise a degree of judgment for themselves. Not only is the kingdom hidden from this kind of wise man, their judgments do the gospel of the kingdom a great deal of disservice by confusing citizens and strangers alike by giving all kinds of reasons why the kingdom of God does not exist and neither does its king!

It is proposed to deal with two schools of intellectual confounders. The first school asserts that some precious teachings in the bible are not literal but belong to the world of mythology. The second school goes beyond what is written by exaggerating the supernatural, such as by telling stories about territorial spirits either on the say so of shamans or by using their own fertile imaginations. The non-literalists regard Adam and Eve, the incarnation of Jesus, and the resurrection as myths. These allege that Peter had come to the end of his tether. He had followed Jesus for three years and then denied him. Jesus had been arrested and executed and he was in fear of arrest. In a state of hysteria he experienced a psychological let out. So great was the power of hysteria within a small community that in the evening in candlelight it seemed as if the Lord came through the locked door to them, and away again.

Further aberrations are that Adam probably never existed, and that the Son of God coming down from heaven being born as a human baby can be seen as "a mythological expression of immense significance of our encounter with one in whose presence we have found ourselves to be at the same time in the presence of God." To which is added the explanation, "that although when in Jesus' presence we should have felt we were in the presence of God – not in the sense that the man Jesus is literally God, but in the sense that he was so totally conscious of God that we could catch something of that consciousness by spiritual contagion." Now all this smacks of convoluted academic jargon.

It is dangerous. When Jesus is not acknowledged as God he is reduced to being a great man who can become a role model for us to follow. In fact he has been compared to Winston Churchill and Mahatma Gandhi. The divine character of Christ is ignored enabling critics to state that biblical fundamentalism has been outgrown. This dispenses with teaching about need for spiritual rebirth and the relevance of the cross. It brigades the standing of Christianity as equal to that of other faiths and humanism. Although the word 'myth' means a fantasy to the normal person, to the non-literalist it means anything that cannot be accepted intellectually. These accuse the Apostle John of putting words on the lips of Jesus thirty years after his death. This they say he did out of desperation to counter the teaching of the Samaritans who saw Jesus as a teacher akin to the Zealots, and not as Jesus claimed, the Son of God. Not everything the non-literalists put about, however, is downgrade but whenever something uplifting is written by them it is usually followed at once by the sort of stuff described here – like nails in a coffin.

For instance, the biblical fact that Jesus heralded the coming of God's kingdom on earth, and that to this extent the emphases of the synoptic gospels were different, is supported by them; only to be spoiled by an assertion that he was not obviously a supernatural visitant. To which they add although the doctrine of the incarnation may have been appropriate to the age in which it arose it should not be treated as an unalterable truth binding upon all future generations. In other respects 'wise' theologians have latched on to loopholes in post-Apostolic writings (e.g. some of The Early Fathers) in order to discredit the Christian message. That is why such writings have been eschewed by the author.

A further example of natural wisdom is the thinking of a modernist Professor. In the process of penetrating some of the weaknesses of Calvanism and Arminianism he came up with something that is far less credible than either. His new set of theories has been called *God's Openness*. The main thrust is a suggestion that God does not know everything – "there are aspects about the future that even God does not know". Divine knowledge is, therefore, exhaustive. How history will go is not a foregone conclusion. Repentance is a metaphor that should not be pressed too far. God knows some things as certain and other things as only possible. His foreknowledge is not exhaustive: the future is not entirely settled. What kind of a God is this! Then follows some contradiction. "God does not promise what he cannot deliver." How can this line up with the statement that God does not know the future? Then there is gobbledegook – "God is temporarily everlasting rather than eternal!" Then something the mind boggles at – "it may even be that in heaven there will be no possibility of falling away!" Not surprisingly, this new 'openness' is admitted to be only a research programme – "it will take some time to mature but a number of 'stimulating ideas' have been raised." That is his opinion about the value of his presentations.

What has been covered so far is a range of academic ideas. As such, they will only circulate gradually into the everyday world, if at all, or else gather dust on shelves, or be sold for a pittance eventually in charity shops. The main danger is that these academic ideas may infect the minds of undergraduates at source. Borrowing a leaf from one of their own books: their ideas may be mentally contagious. One surprisingly paradoxical thing about not a few non-literalists is that they are attracted to sacraments and particularly the doctrine of the real presence. One might expect this is about the last thing they would want to espouse. Eucharistic theology was not fully developed until over a thousand years after Christ. It does not stand up to the fullness of the scriptures, and so rests much more on tradition. A simplistic teaching of The Lord's table or Breaking of Bread fits in much more with The Acts of the Apostles and epistles. One would have thought that intellectual non-literalists would oppose religious ritualism, but, clearly, they prefer to support the use of symbols rather than acknowledge the power of the Holy Spirit. This is pragmatism not wisdom.

As to the second school, this exaggerates the supernatural. It appeals to peoples' emotions, causing them to overlook biblical teaching and to subjugate rational thought to desires of seeing 'things happening'. Novel new ideas appeal to some mainstream Christians. On embracing excesses their inner lives are affected in sinister ways. They do not realise the danger. This second school does not question the literal truths of the scriptures. It seeks to appropriate God's power written in them without waiting upon God's sovereign will, despite the biblical teaching about the folly of such action. There is initial attraction to meetings that generate hysteria and emotional behaviour because the organisers recognise that in the present kingdom there is a battle going on against the powers of darkness. Unfortunately, they over emphasise the role of demons by thinking that the god of this world is always Satan rather than, very often, people sowing to the flesh. They also fail to wait on God and in their impatience develop teachings on the say so of shamans and their own vivid imaginations. A favourite exposition is based on an idea that territorial spirits dominate towns and cities. Apparently, they say, there are grades of spirit with varying degrees of maliciousness. They must not wrestle directly with the devil because he can only be in one location at a time. They are called to wrestle with ground level spirits that are frequently mentioned in the gospels (no references offered). They also identify middle level spirits operating through witches, occult practitioners, and mediums. A spirit of divination is a high level spirit. Casting one out landed Paul and Silas in jail. Wait a minute. Surely it was not so much a girl's demonic gift that got them into trouble but the anger of her manipulators at loosing their source of income? This hits the nail on the head. At every nasty human act they allege that it is demons who are coming out of the woodwork.

Then we come to bizarre anecdotal encounters, all of which caricature and ridicule the legitimate work of the kingdom. A high-ranking occult leader in Port Harcourt, Nigeria, claims that at one time Satan had assigned him the control of 12 spirits and each spirit controlled 600 demons for a total of 7,212. These spirits attached themselves to people, buildings, seats of government and other offices. It is a must to discover the name of a spirit in charge of a town and then engage in spiritual warfare against him. A reservation is that they are not sure whether it is safe to do this as it is alleged a minister in Ghana ordered a devil's tree to be cut down whereupon he fell dead. To the author, who lived in West Africa for years and travelled extensively throughout South and East Africa, it does not seem possible to give credence to stories like this. Witchdoctors have been mentioned in another chapter and instances of evil cited. It is a fact that when Costains (an international civil engineering company) had to repair a culvert in the Cameroons it kept falling in quite inexplicably. It was only when engineers bought a crate of whiskey for the local elders, on a pretext it was for a pouring a libation, that they succeeded in making things good. It was established that the elders drank most of the whiskey. However, there was never any evidence that witchdoctors ever had power over white people. This suggests that the indigenous population had a shocking fear complex.

A set of naturally wise theologians and writers allege that there are spirits of fear, lust, pride, apathy, witchcraft, pleasure, and religion. Experience has indicated that once the idea of these spirits is spread, unstable people begin to imagine they discern these. Unprincipled power seekers use this knowledge to attack their rivals. Several ministers have been driven out of churches by ambitious deacons who accuse them of having a spirit of witchcraft. In one instance this was interpreted as meaning the lobbying members for support. Another deacon alleged that he discerned a spirit of lust in a lady member as she was singing the closing hymn. A newly converted couple in their fifties were told in their house group that the husband had inherited a curse from his father which had to be cast out. This was after he had mentioned that this father had abandoned his mother when he was a small child. A well qualified Christian doctor was persuaded he had inherited a curse because his grandfather had been a Freemason. Once people were induced into thinking that they could intuit angels there were numerous claims of sightings of angels. The disappearances of planes and ships in the Bermuda Triangle have been attributed by these territorial spirit people to the machinations of the spirits of slaves who were thrown overboard. Many sound Christians have taken part in 'Marches for Jesus' because they were oblivious to the fact that the organisers were territorial spirit merchants who were marching round in opposition to the town's head demon.

When the Vineyard Fellowship in Illinois prayed and fasted it claimed that a grotesque unnatural being appeared and growled, "Why are you bothering me?" It claimed to be a demon of witchcraft who had authority over the geographical area. When the streets of some cities were named to him he is reported as saying, "I don't want to give you that much." In the name of Jesus he was ordered to give up the territory. The church then doubled from 75 to 150 with most of the new entrants coming from witchcraft. It was claimed that nearly all new believers had to be delivered from demons. There is only one thing to do about all this and that is to use the word of God. We must cast down imaginations and every high thing that exalts itself against the knowledge of God, and bring into captivity every thought to the obedience of Christ. The emphasis here is on "imaginations". Just look at Paul's concern for the Galatians. Before they knew God they were slaves to those who by nature are not gods. So how was it that now that they knew God they were turning back to those weak and miserable principles. The problem was the influence of infiltrators who were zealous (or 'wise'?) to win them over. But for no good. What they wanted was to alienate them from Paul and to make them zealous for them. They were false teachers and sheep stealers. They had got the concept of the operation of the kingdom completely out of balance and they were spoiling it for others.

The fact that true spirituality could be hidden from the wise, as Jesus said on a particular occasion that it had been, means that such a blockage causes some to resort to other sources of revelation. The examples given in this chapter of various teachings were sourced either from natural reasoning, brain-storming, or, in some cases, by evil influences. In is not hard to see the cunning behind the latter. People of the first school are deterred from seeking the real power of God, and of the second by eventually being dissatisfied with counterfeit excess. The bible states that the folly of people will be manifested. It is interesting that when the children of Israel began to loath the manna that was sent from heaven, God sent then quails that choked their nostrils.

The Apostle Paul gave a masterly view of innate wisdom. He wrote that the message of the cross was foolishness to those who are perishing, but to those who are being saved it is the power of God. For it was written , "I will destroy the wisdom of the wise; the intelligence of the intelligent I will frustrate." Where is the wise man? Where is the scholar? Where is the philosopher of this age? Has not God made foolish the wisdom of this world? For in the wisdom of God the world through its wisdom did not know him. God was pleased through the foolishness of what was preached to save those who believe . . . the foolishness of God is wiser than man's wisdom, and the weakness of God is stronger than man's strength. The Apostle James wrote that there were two kinds of wisdom, but that only one kind came from above. In fact he confirmed what has been explained elsewhere that it is the fruit of the spirit that is paramount. True wisdom is to show by good deeds and humility what acceptable life in the kingdom consists of. This is as opposed to harbouring bitter envy and selfish ambition in the heart. This latter kind of "wisdom" does not come down from heaven but is earthly, unspiritual, of the devil. But the wisdom that comes from heaven is pure, peace-loving, full of mercy and good fruit, impartial, and sincere. Peacemakers who sow in peace reap a harvest of righteousness. And such is the kingdom of heaven.

14 | Testimonies of the Kingdom

"Now I'm trusting every moment,
Less than this is not enough;
And my Saviour bears me gently,
O'er the places once so rough."
                        F.A. Blackmer

"The Lord shall preserve thy going out and thy coming in from this time forth and for evermore." *Psalm 121:8*

There are instances in any trusting citizen's life which can be attributed to help from a divine source. These range from a public attested miracle, as when Peter was released from chains in prison by an angel and then the doors were miraculously opened without warders being aware of it, to innumerable cases of a much humbler sort. These might very well help ordinary people and, therefore, may strike readers as being more relevant to their own lives. A young soldier with a religious background was conscripted into the army in World War II and posted as a trained signalman to an infantry regiment stationed in Northern Ireland. A typical sentence in the London courts was – six months or join that regiment.

It was very tough going both as to the background of fellow privates and the rigorous training programme in the Sperrin mountains. Exercises involved deliberately organised hardships such as going without normal rations, the use of live ammunition resulting in casualties, and crossing the river Bann on a moonless night in flimsy collapsible boats when equipped with a full field service marching order including two pouches of ammunition. At least twenty-eight men drowned. Some thought this was all worth enduring if it would help to win the war.

Having been keyed up for an overseas draft there came a day of anti-climax when the Commanding Officer called the full regiment on parade. Surely, this was to announce going into action. He announced that someone had been using his private toilet and if this continued he would start using theirs. His actual words were far too racy to bear repetition here or anywhere else. After enduring the coarseness of his companions, the hard military life, and then having this anti-climax, the young soldier went out and cried to the Lord about the situation he found himself in. He was under the impression that God spoke to him and told him that he would be out of the army in five weeks. He doubted the promise. On return to camp he saw a notice calling for volunteers for signallers in The Royal Navy and put his name forward, but so did all the other signalmen in the specialist platoon. It was decided to conduct tests as a means of releasing the three who performed best. The young soldier came top. As he was the commanding officer's signalman the idea was abandoned and the men told to cut playing cards. The young soldier drew an ace. In exactly five weeks he was transferred to the Navy. To put a peg into this account one of the three became member of Parliament for Peckham for many years and was Private Parliamentary Secretary to Lord Dennis Healey who was then Minister of State for Defence.

The same young soldier got into trouble on a route marching exercise. As a signalman he carried a heavy wireless set and was given a Sten gun. Other members of the platoon carried a mortar and bombs and ordinary rifles. The procedure was to march for an hour and rest for ten minutes and so on. At each halt heavy equipment was changed from one man to another to share the burden. Because of this arrangement a soldier who normally carried a rifle took over the wireless set and Sten gun. At the next stop these were passed to another. When the time came to return them to the original owner it was apparent that the Sten gun had been left by the roadside. The blame, quite unfairly, fell on the signalman and he was subjected to jibes of how long he would be sent to prison for, especially as the IRA had probably gained the weapon. He could only pray about the matter. On return to camp the company sergeant major suggested that D Company might help. The sergeant major of D Company handed over a replacement Sten gun without any quibbles. After having been in the navy for a while the former soldier heard that the army camp had been taken over by another regiment. An order was given to dig slit trenches so that the soldiers had something to do. They dug up a lot of rifles and Sten guns. There had been an inventory error and to cover this up it had been considered necessary to bury the surplus items. It was the D Company sergeant major who was taken to a court martial. Regimental history revealed years after that the Colonel and his wireless operator were killed by a mortar bomb in the Italian campaign. The Sperrin mountains had been chosen as being similar to Italy.

At the close of the war the destroyer the signalman was serving on was chosen for a suicide mission against the Japanese mainland. While on the way the ship was ordered to stand off and the atomic bombs were dropped to put an end to the war.

In East London in the early 1960s a newly converted couple were anxious to spread the gospel message of the kingdom but at that time were not well versed in the word. They took a tape recorded message to the houses of any who were prepared to listen and to marked effect. One of these visits was to a very elderly lady and elderly daughter. There was a newspaper instead of a lampshade in the hall and this showered dust on tall peoples' heads. A permanently set tea table had a bowl of grey sugar. It transpired that the upstairs was let to a niece and her husband. The niece was invited down. It can be said that she was dressed and made up like a young "sixties" person and appeared an unlikely listener. After the tape had been running a few minutes green mascara was running down her face mingled with tears. As it was customary in those parts to go out to the front at church as a witness to salvation the husband was asked to drive her to the chapel on Sunday evening. As it was some distance the husband decided to go in and it finished up that they both went out to the front. The niece's father was a transport manager, the niece was his secretary and the husband was a driver for the same company. That weekend the father went on the run because he was involved in some kind of dishonesty. The two directors suspected the couple of collusion, as it turned out quite unjustifiably. The couple were summoned for an interview on the following morning. They began by relating how they had come to the Lord the previous day. The two directors turned out to be committed Christians and offered the couple a new start in Manchester with accommodation provided. These experiences, excepting that of Peter's, are about the little lives of ordinary people who are precious in God's sight.

It will be appropriate to close this chapter with an incident that was a microcosm of the Lord's mercy in South Africa when a bloodbath was averted by the goodness displayed by those who forgave and the repentance of those who were willing to change. In all ages the purposes of God must be brought about. Even out of evil good can come. In May 1986 the political situation in South Africa was akin to a tinder box. The ANC had large armies of revolutionaries over various borders, especially in Zambia, and there was unrest, repression and bloodshed in the country itself. A President of an international professional chartered institute was making an official visit to Johannesburg and Port Elizabeth with his wife. The latter's diary records. "I met the mayor and he said he thanked God that my husband and I were Christians. He then said he would arrange for an inter-denominational prayer meeting in the Mayor's reception room. He was encouraged by the fervour of the prayers and the response of "amens". Although he was Dutch Reform he said he felt like joining in the responses. I talked with the mayor's secretary and when I related my conversation with the mayor her eyes lit up. She said she would arrange a day of Prayer. I said we had prayed a lot about the tour. The mayor then convened a joint public prayer meeting for Black, White, and coloured. His invitation read as follows: With the deepest conviction that the future of our country rests entirely with Our Lord, and with faith in his promise given in II Chronicles 7.14 that our sins will be forgiven and our land healed, if our people humble themselves and pray, I intend to declare Sunday, 22 June 1986 a day of prayer for all God-fearing citizens of Port Elizabeth and to organize an interdenominational prayer meeting on 19 May at 1500 in the City Hall Auditorium". The resolution of South Africa's problems was little short of miraculous and the attitude in Port Elizabeth contributed in a small way to a healing and reconciliation that went across the nation. Nelson Mandela was greatly used by God to accomplish His purposes with a minimum of trouble.

The President's tour continued to Zambia and countries farther North. On the night of arrival in Lusaka the South African Apartheid Government bombed the ANC camps near the capital. President Dr. Kenneth Kaunda, a Christian, showed no racist bias and entertained the visitor at his palace and joined in a television programme with him. The Zambian President mentioned the unequal yoke between large and small countries. His father had been a Pastor and taught him this. The reply of the guest was that he was a Christian and that Jesus said his yoke was easy and his burden light. Doctor Kaunda was referring to the unequal yoke in 2 Corinthians 6.14 that is about association with the world. The thought he had was that grants from the rich nations had strings attached to them. What Jesus was referring to in Mark 11.40 was that he would be directing the oxen. He was thinking of an opposite situation. With two kindred people yoked together the burden is shared. A further thought based on Philippians 4:3 is that Christians are loyal yoke-fellows. There was then extended an invitation to lunch at the palace the following day. Since then the kingdom has grown enormously in West, South and East Africa. In Zambia the following President Mr. Chilombie dedicated the whole country to the service of Jesus Christ.

Identities have been kept out of the narrative for the most part including the author's. It is necessary at this juncture, however, to explain how this book came to be written and why the author was qualified to write it. In an Orthodox upbringing, unwavering admiration and love for the persons of Jesus was evidenced from the age of four; as a pupil there was an award of an Archbishop's Medal for Religious Knowledge, and as a student winning of a first place in a national examination in Eucharistic studies. The invigilator, Monsignor Valentine Elwes, wrote, "The standard of answers would have done credit to someone studying for the priesthood." There followed an invitation to embark from Father Patrick O' Connell (subsequently Senior Chaplain to the RAF, a CBE, and Bishop of Southampton). The author was conscripted into the army in 1942, serving in Northern Ireland, and then transferred to the Royal Navy. He served on the same flagship as Valentine Elwes. (It is of historical interest that the latter had served in the First World War as a Lieutenant Commander, was axed under The Geddes Act in 1923, and followed a late vocation in Rome, becoming a Papal Prelate. On rejoining the navy at the outbreak of the Second World War he became a Senior Chaplain. In 1945 he went ashore in Yokohama, devoting his life to helping the Japanese).

After the war the author could no longer accept the concept of extra-biblical tradition, and the rigid authoritarianism of Orthodoxy. He underwent a new religious experience. Interested readers can read John's Gospel, chapter 6, 63, and Psalm 130, in order to understand why there was a prompting to call on The Lord. He joined the Free Church without animosity or recrimination. He, thereafter, relied upon lengthy studies of scripture. In 1968 the Principal of Redcliffe Missionary Training College, Miss Nora Vicars asked the author to give a series of lectures, each year, to the whole college on Contemporary Religion. Students were already qualified professionals (e.g. doctors, nurses, teachers). Most countries would only admit Christians who could offer a scarce secular service. The purpose of the lectures was to familiarise students in advance with the beliefs of missionaries of other denominations, and of indigenous populations. The lectures were repeated publicly in Southend, Essex in 1983, and Marlow, Buckinghamshire, in 1993. The material form the rump of Part One.

Dr. D. Martyn Lloyd-Jones read the lectures in 1975, admired them and urged that these should be made into a book. "The Doctor" died in 1981, but a manuscript based on the lectures was approved by his daughter, Lady Catherwood. This was not published. The author's association with "The Doctor" from 1966 was close. As examples: joint convenors of conferences at Cloverly Hall, Shropshire; coinciding ministry on The Isle of Wight, and a pastoral assignment by him in

Accrington. Most importantly, there were opportunities for long personal conversations, followed by correspondence.

Due to "The Doctors" graciousness, the different backgrounds of Welsh Calvinist Methodism on his part, and Orthodoxy did not hinder a unanimity in Christ. This meant agreement about:

1. A present dynamic spiritual Kingdom of God on earch evidenced by the Lord's Prayer.
2. Receipt of the life of Christ within the heart conveyed by the power of the Holy Spirit.
3. Subsequent empowerment of the Holy Spirit for Kingdom service.
4. An authentic Christian must bear fruit. Empty profession being no better than a ritualistic ceremony unaccompanied by the power of the spirit.
5. The biblical meaning of the word 'church' applied only to Authentic Christians.

It is important to note that the author did not discuss the issues of "election", "pre-destination" and eschatology. In his published lectures on *What is the Church?* (July 1969, Evangelical Press), "the Doctor made his views absolutely clear. He did say, "I am not going to deal with the question of the visible and invisible church."

15 | Prayer and the Kingdom

"The effectual fervent prayer of a righteous man availed much."
James 5:16

When Jesus was asked by his followers to teach them to pray he responded by framing a model example for their use. Now known as The Lord's Prayer, The Children's Prayer, or The Our Father, it takes on a deeper meaning if said in the realisation that Jesus had in mind the invisible earthly spiritual kingdom that is in the here and now. Only a sublime divine mind could utter a prayer like this so spontaneously and succinctly.

"Our Father who art in Heaven" There was no formal teaching about the fatherhood of God. Jesus just spoke about his father easily and as a matter of course as an established fact. His reference to "our" Father associated all of his followers as members of the divine family and children of God. As Paul wrote later, "Because we are sons (and daughters) God has set in our hearts the spirit of his Son, whereby we call him Abba – Father." This process was by adoption.

"Hallowed be Thy name" Those in the kingdom were shown that the Father is to be revered, honoured, and worshipped as holy, just and righteous, being omniscient, omnipotent, and omnipresent.

"Thy Kingdom Come" At the point of time when Jesus gave us this prayer the kingdom on earth was only at hand, although its powers were operative in Jesus and in those to whom he gave authority. A plea for the kingdom to come was literal because at that time it was still

imminent. The words took on a different meaning when the kingdom actually came upon the death and resurrection of The Lord Jesus. The prayer then became one seeking the progress of this kingdom both in the world at large and in more fullness in the hearts and the lives of citizens at large. It still requires that more territory should be invaded that is still in darkness and under domination of the prince of this world. Jesus was against repetitive prayers and this is understandable because the repeated utterance of this one has taken the shine off what it was meant to achieve in spirit.

"Thy will be done on earth as it is in heaven" The kingdom is everywhere where obedience to the righteous rule of God is in evidence: in other words where his will is done. On heavenly territory His rule is complete and established forever. This part of the prayer asks for a similar holistic situation to come about on earth as it prevails in heaven. It recognises the dynamic battle that goes on between the enclave of the kingdom on earth and the kingdoms of this world of which Satan is prince. If the prayer is fully answered it will lead to the conversion of the nations and a total victory of light over darkness. Even if it is not fully realised it does permit a more optimistic slant to the end times rather than a completely gloomy one.

"Give us this day our daily bread" We need natural sustenance, as recognised by Jesus in the multiplication of the loaves. We need deliverance from famine: a much greater reality in the Third World. The Israelites were fed in the wilderness by God sending down manna from heaven. As a simile Jesus said that the bread he would give would lead to eternal life. When asked that he should evermore give this bread he said, "I am the bread of life. He who comes to me will never go hungry."

"And forgive us our trespasses" This does not refer to initial forgiveness because children of the kingdom (citizens) are already justified and cleansed. However, if we say we have no sin we deceive ourselves. It is scriptural that we need to confess our sins to God. If we confess our sins he is faithful and just to forgive us our sins and purify us from all unrighteousness.

"As we forgive them that trespass against us" We are to go on forgiving others indefinitely, expressed as "seventy times seven". It can be taken that this is conditional upon contrition as God forgives us only after repentance. God's provisional forgiveness is boundless because while we were yet sinners Christ died for us. The question of forgiveness and judgment will be debated when eschatology is covered.

"And lead us not into temptation" This has caused a lot of difficulty because God leads no one into temptation. One must be honest and admit it is not possible to be certain about the exact meaning. It has been suggested that two petitions should be run rapidly into each other: lead us not into temptation but deliver us from evil. This is based on the fact that it is more likely to have meant a plea that we should be kept from temptation or situations that are dangerous to our souls. It is Satan who tries to lead us into temptation as he did in the wilderness when Jesus was fasting.

"But deliver us from evil" This is a prayer for deliverance from the snares of Satan, from the seductions of this world, and from the propensities of our sinful natures. It also refers to all the evil that wars against citizens in their work for the present kingdom.

"For Thine is the kingdom, the power, and the glory for ever and ever. Amen" The Father's kingdom had no beginning and it will have no end. Its capital is the New Jerusalem. Its future stages will be under discussion in Part Two where a line is drawn between the periods when Jesus will reign and when he will hand over the kingdom to his Father. A catechistical definition of prayer is that it is a raising up of the mind and heart to God. Before he ascended Jesus gave the Great Commission to his followers to go into all the world to preach the gospel. He also told them to tarry in Jerusalem to await the promise of the Father – the gift of the Holy Spirit. After this they continued for days in prayer and supplication. How important it is that his people continue to pray that his will shall be done on earth! Thy kingdom come! This applies to one's own needs – give us this day our daily bread! The "our" also applies to the needs of others. Pray for one another that you may be healed.

Another great prayer was when Jesus prayed to his Father just before he left the world. He said the father had given him authority over all people, and to give eternal life to those who received him. He had completed his work. He was not praying for the world but for those whom the Father had given him. He was going back to the Father. He asked for these people a full measure of joy; that they might be protected from evil, and sanctified by the truth of the word. The glory the father had given him he had given them that they might be one. "I in them and you in me." How true this is! And how marvellously was this prayer confirmed. Paul wrote of this later as the mystery hid from the ages which is: "Christ in us the hope of glory." Another important prayer is the prayer of faith mentioned by James.

If anyone was sick, they should send for the elders of the church to pray and anoint with oil. The prayer of faith would raise the person up and if they had sinned there would be forgiveness. In the permissive will of God this circumvents adverse affects of sickness. It cannot nullify the overall penalty of Adam. Lazarus died eventually. Despite the basic example of the Lord's prayer we do not always know what to pray for as we ought. This is where the Holy Spirit steps in. He helps us in our weakness, intercedes with groans that words cannot express. He searches our hearts and knows the mind of the spirit for the spirit intercedes for the saints (citizens) in accordance with God's will.

In all cases of known answers to prayer, care should be taken to give thanks and remain permanently thankful. Wisdom must be exercised as to when it is advisable to tell others about answers to prayer. A line should be drawn between the possibility of harmful criticism and the likelihood of depriving others of rejoicing. It will not be known to what extent prayers were answered this side of glory. It is likely that many visible blessings, revivals, deliverances and unknown avoidances of disasters were due to answers to prayer. Anyone can call on God but the prayers of his own are particularly precious in his sight. God can speak to us through the spirit more readily when we draw aside to a quiet place. "When you pray do not stand at the street corners or stand in the synagogue – go into your room and shut the door and pray in secret; and the Father which says in secret will reward you openly" (Matthew 6. 6). Peter went onto the roof of his house where God gave him an all important vision that the kingdom was open to the gentiles. He also indicated in advance that Cornelius was persona grata. Therefore the double advice is that we should go into our room and close the door fast. Firstly, our praying should not be ostentatious, and secondly it should be in a private place in which to pray before God. It is not known to what extent communication between Jesus and his father was two-way or whether these were audible or by the spirit. It is more appropriate for us to pray audibly either when alone or in a prayer meeting. Otherwise mental prayer can be resorted to. We pray for others, we pray for ourselves, and we pray into situations. With God's help and by his grace may we realise with greater force a need to relate The Lord's Prayer to the present kingdom battle against the powers of darkness. "Thy kingdom come. Thy will be done on earth as it is in heaven . . . deliver us from evil. For Thine is the kingdom, the power and the glory . . . for ever and ever Amen."

16 | Denouement

'If seven maids with seven mops
Swept it for half a year,
Do you suppose,' the Walrus said,
'That they could get it clear?'
'I doubt it,' said the Carpenter,
'And shed a bitter tear.'
Through the Looking-Glass, Lewis Carroll (Charles Lutwidge Dodgson)

In Association Football one of the roles of a mid-field player is to act as what is termed a sweeper. Since he plays mainly in a part of the pitch where attacks can be initiated by one side or the other, or the ball can go loose; potentially it can be "swept up". That means it can be dealt with effectively and an advance be made towards the forwards or to the goal. Some balls go into touch (i.e. out of play) despite the efforts of players. This is an excellent analogy as to how one can deal with those aspects of scripture that either require more erudite explanations or others that are beyond the comprehension of people so far. So in this chapter some explanations of difficult scriptures are attempted, some issues are confessed as beyond finding out, and some are brigaded as being kept from us for divine reasons. This is not meant to be a chapter of doubts but a channel for discussing problematic aspects of scripture in a frank and honest manner. What can be swept up will be dealt with first.

In the preceding chapters it was shown how God's hand had been instrumental in bringing his kingdom into the history of this world two millennia ago. Something of its nature and modus operandi was described together with powers that oppose it. It remains here to unravel a number of complexities before passing on to Part Two in which biblical predictions about how history will be superseded by the one eternal everlasting kingdom are set out. It is certain that while time lasts Satan's kingdoms of this world will continue in their hostility against righteousness. What is uncertain is the extent to which the true kingdom will grow on earth prior to the end times. The Holy Spirit will continue to convict the world of sin and the gospel will be taken to the whole world. Believers will continue to swell the eternal sector of the kingdom through death. Unknown factors are the actual degree to which the nations will respond and the numbers who will work actively in support of the kingdom. For instance, will the final great prayer of Jesus, "that they may all be one", be answered sooner or later? His petition could be given greater impetus if churches reviewed their priorities in the light of scripture.

Before discussing the possibilities of unity and some of the barriers hindering it, a few problems are aired. Some have been perplexed by a seemingly hard statement by Jesus that we should hate father and mother. We should be grateful for the richness of our English vocabulary. In Aramaic, the language Jesus spoke, there are no words between love and hate giving intermediate meanings. This would have been understood at the time. Jesus meant that we should love God even more than we love our parents.

It may be wondered why God did not reveal to Jesus exactly when the second coming would occur and, therefore, when the invisible kingdom would become a visible reality. During the Second World War England had to be blacked out by night; car headlights were reduced to slits with shaded vents, and all road signs were removed. This was inconvenient and dangerous but it was preferable to allowing the enemy to bomb more accurately. In the event of invasion the enemy would be frustrated in finding its way. Likewise the devil would dearly like to know the time of his final defeat. He could then plan for it. Currently he is confined to devices limited to his own cunning and to cheap jibes – "Why don't you come down from the cross and save yourself as you saved others?" A ploy from his point of view but utterly fruitless. Then – "Where are the signs of his coming?" Scoffers will come saying, "Where is this 'coming' he promised? Even since our fathers died, everything goes on as it has since the beginning of creation." This is a satanically inspired deceit at complete odds with biblical advice to watch and pray, and a warning that the second coming of Jesus will be as a thief in the night – that is as to lack of early warning. Yes, it is true that as they behaved in the days of Noah and of Lot, people will be similarly occupied at the end, but the bible now warns of its inevitability. Jesus was warning of the dangers of complacency.

Then there is the proverbial question as to why God permits evil and suffering. People tend to avoid saying that God is directly responsible but are perplexed why he does not prevent tragedy. There is a feasible explanation in that when God as a spirit dwelt in eternity alone he desired spontaneous companionship with others. To this end he first created angels. As to angels and, later, created human beings, it would not give much satisfaction if these had not been given free will. It is thought that angels had a one-off opportunity and that is when some chose to rebel. If beings were like puppets on a string – what satisfaction would that give to God and others? Nor would it do to treat these like domestic animals! So the blame for evil clearly does not rest with God but mainly with Satan. Was he the head of a pre-Adamic race? That is pure conjecture. It is known that he deceived Adam and Eve. It is in the munificence and wisdom of God that he was able to turn what seemed a mistake into a new creation: something that was infinitely better. When permitting choice God was aware through foreknowledge how people would react. This would be for better or for worse but he knew his decision would be worth it in the end. Critics should reflect, when they are peeved, that without God there would be no reason for them to exist at all. Evil will be turned into good step by counter step. God does not will that any should be lost but will not prevent them from self-destruction. Conveniently, all the evil perpetrated by the devil and his deliberate and deluded helpers is turned by God into a testing ground and means of pruning of those who enter the kingdom. Those who do not accept this hypothesis fail to appreciate the full glories of the eternal abode planned by the love of our creator.

Then there are imponderables. The redemption of all creation will be discussed in Part Two but here the present standing of animals will be raised. In the beginning animals were not eaten by other animals or by Adam and Eve. All fed on grass or fruit. Adam was a horticulturalist and his clothing was fig leaves. That is not skins or furs of animals. After the redemption of all things this will again be the case. It will be a perfect situation. The wolf will lay down with the lamb. After creation was marred God used an animal to provide Adam and Eve with covering. This was an accommodation in sadly changed circumstances. A food chain came into being. In the same way the beauty of some spectacular scenery was only a pale reflection of the world when God observed that the creation was good. Abel sacrificed an animal and Noah offered sacrifices of animals and birds, and "God was pleased with the aroma". In a covenant God gave to Noah everything that lives and moves to be eaten just as he had formerly given green plants. This seems very hard on the animals. It is clear that the sacrifice of animals under Judaism was a type or shadow of the one great sacrifice of Jesus on the cross but not clear why this was necessary as a regular Old Testament ceremony. It is, however, abundantly clear that the cross of Jesus is fully explicable as in the purpose of a holy God in order to make forgiveness possible and all that followed after it in resurrection and eternal life. The sacrificial system ended precipitately and Judaism became a bloodless religion. Jesus said God would have obedience and not sacrifice. So why should there have been the slaughter of countless animals and why should this please God? Why did Jesus eat lamb and fish? It could be that fallen creation and fallen nature had to be lived with. Jesus emptied himself of an equality with God and took upon himself the form of a servant. It is a deep and wonderful fact that Jesus shared our humanity. The extent to which God is prepared to come alongside humanity, to endure its contradictions, in order to ensure deliverance, is not something that can be fully comprehended and this should be admitted. If Rugby Union Football was taken as an analogy instead of Association Football then every ball would be kicked intentionally into touch and thrown back in to be fought over. This is particularly symptomatic of the Liberal Church which pulls scripture to pieces and partly applicable to the behaviour of some denominationalists – at least as to the scrum down rule.

It is sometimes wondered why the angels of God could not escape supernaturally at Sodom and that it became necessary to surrender women to the mob in order to save them. One explanation is that "angels" are not always angelic beings but humans used by God as messengers or helpers. Certainly this would gel with a New testament instruction to beware "lest you are entertaining angels unaware".

That God gave Moses the ten commandments is obvious from their very contents. As to the law it is interesting that Jesus attributed this to Moses and it could be that as sound as many of the principles were and that Moses was a God-centred vessel these could have been at least partially subject to the spirit of Moses. The Jews were inclined to give Moses credit for everything. There may have been elements of the law that did originate from Moses by delegation. What is certain is that any power experienced was rooted in the finger of God. Even the Lord Jesus himself saw that he only did what his father showed him. Jesus had to remind the Jews that it was his father and not Moses who gave the bread from heaven – adding, " I am the living bread that came down from heaven. If anyone eats this bread they will never die." This had an entirely spiritual connotation. Moses was the author of the Pentateuch and according to the Book of Deuteronomy unless an animal chewed the cud it was considered unclean. It was tradition that Jews did not fraternise with gentiles. God revealed to Peter in a vision that Gentiles were to be treated equally under the new covenant. He did so by showing that anything God regarded as clean could be eaten including creatures formerly prohibited under the law of Moses. So some of the law might have been based on health considerations rather than being on divine tablets of stone?

Whenever problems are faced and possible solutions offered it is a good thing to bear in mind a wonderful chapter in the Book of Job. This puts all of our puny ideas into perspective and raises uncertainty about the wisdom of querying divine things. One good rule is to heed the instruction not to go beyond what is written but this is what many ignore in relation to last things, as we shall see. We can only give a view and make it clear that this is simply an attempt to fathom uncertain matters without claiming infallibility or disparaging the views of others unnecessarily. God acknowledged that Job was a good man so he wondered why God was allowing his distress. Although this is not a popular view, he may have been self-righteous. God asked Job where he was when He laid the foundations of the earth, and recounted many instances of creation. For instance, could he bind the sweet influences of Pleiades or loose the bands of Orion? Where was he when the morning stars sang together, and the sons of God shouted for joy? Job could not answer. " What shall I answer you? I will lay my hand on my mouth."

Unity within Christendom is regarded as a highly desirable but elusive goal. This subject will be returned to in A Review: Mystery, Majesty, Immanence, Chapter 23 of Part Three. If there was a universal recognition of a present invisible sector of the kingdom on earth and more of our attention was given to this the result could be harmony and unity where at present there is conflict and division. At the present time denominations are too concerned with fighting their own corners. Rather than defending the faith once delivered to the saints they defend their own versions of it. There are such large differences of approach that an impartial bystander would think this to be strange as there can only be one fundamental truth. If there was a standing back and reflecting, as in Ezra's day, it would quite soon become apparent that they could not differ and still all be right. No church can be exempted from blame for failures even if it is only because of unwillingness to be involved in positive initiatives. The bible states that sometimes there has to be division in order to establish who is right. The contretemps between Paul on the one hand and Peter and James on the other is an example. And so is the spirit in which it was resolved! The following discussion is not intended to be seen as a means of indictment. It is meant to identify backgrounds that render any possibility of close unity to be highly unlikely. However with God all things are possible.

Firstly, Orthodoxy, numerically the largest wing of Christendom, has had a tempestuous history. It has been the subject of internal divisions. The Western and Eastern churches split on the *filioque procedit* controversy. Did the Holy Spirit proceed from the Father and the Son? The suspicion is that this argument was used in what was really a power struggle. Orthodoxy has built an edifice of self-justification by teaching an apostolic succession, a primacy of Peter, and unique sacramental rites. This in its reckoning means that unity can only be brought about by submission of all others to these doctrines. There are grounds for optimism over a reconciliation within Orthodoxy. As to a wider unity, the Roman Catholic Church has changed a lot in its attitude since The Second Vatican Council of 1965. It has pursued a benign ecumenical policy and does not criticise other religions as it used to. Nevertheless its teaching of Papal Infallibility since 1875 constrains it from any rapprochements. The victory of the Ultramontanes over Gallicanism did not help the cause of unity. The highly respected Cardinal Hume could only offer Anglican dissident clergy "the full menu" because of the system. The equally respected Cardinal Newman could not accept the doctrine of Transubstantiation before he went over to Rome. He then submitted to it on grounds that they must know better. Since the Nicene Creed, which most Christians would accept, the Roman Church has added a list of compulsory beliefs, mostly on grounds of tradition and not the bible. It seems a great pity that it has bound its adherents to these beliefs on threat of excommunication although this is back-pedalled in the light of increasing defections and intellectual misgivings.

Despite its liberation from the persecution of the Communists the Russian and Eastern Orthodox churches are very antagonistic towards evangelical Christianity, very unkindly labelling it as a cult. It would be a great benefit if Orthodoxy in general started to allow freedom of conscience on matters that have been added to scripture. It would be difficult for it to accept overtly a doctrine of a present kingdom with entry by spirit baptism unless it modified teaching on regeneration by water baptismal rites alone. The words, "I make you a child of God and an inheritor of the kingdom of heaven" must be accompanied by the power of the Holy Spirit, causing an appropriate change of life. Otherwise, there is great danger of promoting a sense of false security. The only unity spoken of by Jesus was "I (Jesus) in them and you (Father) in me that they may be one as we are one". That is by a process of new birth. Except people are born again of water and the spirit, water being a term for the word, they cannot enter the Kingdom of God. Paul wrote about the mystery hidden from the ages, "which is Christ in us the hope of Glory." This is linked to the present kingdom because it is by new birth that people gain admittance to it. Sacerdotal rituals symbolise spiritual realities but just as the law was a schoolmaster that brought Paul to Christ so people need to see in symbolism a pointer to new life in a risen saviour. An acceptance of kingdom teaching would avoid syncretism, which seems to be the goal of some ecumenists.

Roman Catholics now regard Anglicans as "separated brethren". They pray that together with them they may be united in one fold. This unity is under the Shepherd and Bishop of souls. This shepherd is the Roman Pontiff. This would not be unity but submission.

Secondly, as far as Protestantism was concerned it had a sound inheritance. Its origins were grounded in a reversion to the early church's basic doctrines. By protesting against the mercenary corruption of indulgences and added ritualistic practices it had a good start as a champion of truth. As the Reformation gained momentum Protestantism became tainted. Apart from doctrinal embellishments it developed its own factions. Even over those matters in which it was right it became belligerent and then it resorted to false scandal rousing something that was entirely unnecessary. The exaggerations of Hislop in his "Two Babylons" have been mentioned elsewhere. Two further examples are H.G. Wells' *Crux Ansata* and *The History of Maria Monk*. The former only served to antagonise Catholics when a reasonable critique might have met with interest. It was confined entirely to linking the philosophies of Adolph Hitler and Benito Mussolini with Roman Catholicism just because their parents had them baptised into that church as infants. The latter consisted of scurrilous and totally untrue stories of a women who alleged she had been abused by priests in a convent in Canada. It was read very widely and was circulating amongst soldiers. A catholic soldier reported this to his uncle who researched the book and obtained official acknowledgment of its falsity. This finding was published as *The True History of Maria Monk*. The Oxford Movement and the growth of a "High Church" in Anglicanism merely led to an Internal Roman Rites faction. Liberalism spread from Germany throughout the non-Catholic church and into many university faculties of Theology. The divinity of Christ was denied and so was a literal resurrection. This was a unity of disbelief! It is not normal for Protestants to pray for Catholics in a corporate sense because Rome is held to be a completely false system. Such a view was reciprocated by Catholics until about the middle of the twentieth century.

Attitudes generally are much more circumspect at the present time and have obviously been influenced by rising standards of education and by a growing forum of democratic commentators. However, there is something of a vacuum. Ecumenism seems to be concentrated upon a mixture of syncretism on the one hand and obduracy on the other. Nobody is looking for acceptable common ground in the gospel of the kingdom of God. So far in this chapter some unsatisfactory aspects of organised religion have been described in order to demonstrate a sad lack of unity of purpose over the centuries. Also some minor difficulties over scripture have been addressed. All these matters needed "tidying up" and do not seem to be critical to the resolution of the main purposes of God.

What is of major importance to God are the attitudes of hearts and minds. It will become apparent later just how attitudes after God's own heart will become perfected in his saints. It will suffice here to cite an admirable example. Paul became estranged from the religion of his birth and of blood relationship. Once he had followed the risen Lord after a dramatic meeting on the Damascus Road he was cruelly treated by his former co-religionists. He endured beatings and a stoning. Yet he said that he had great sorrow and unceasing anguish in his heart, and could wish that he was cursed and cut off from Christ for their sake.

On the other hand Moses, who was probably the greatest of God's Old Testament servants, was not allowed to enter the promised land because he criticised his people. He only viewed the land from the top of Pisgah. We might be inclined to sympathise with Moses because he had a lot to put up with. But this was not up to God's expectations. Jesus suffered the contradiction of sinners against himself. He endured the cross because of the great joy that was set before him. He became the perfecter of our faith. Furthermore, from the cross he said, "Father forgive them." "Lay not this sin to their charge."

The difference between the progeny of the first Adam (of the earth earthy) and the second Adam (The Lord from heaven) is a sinful nature in the former as opposed to a divine nature in The Lord. As to the former it can be said that all have sinned and come short of the glory of God. Nobody is good save God. Even those who are born again are still lumbered with a conflicting old nature. That is why in this life there are remaining impediments to fulfilling God's full purposes. This is a remarkable fact and is a vital clue to the manner in which God's kingdom will be brought to a final consummation.

The bible predicts a great apostasy and the judgment will reveal just who the apostates are. It is possible these will be defectors to secularism from all sections of the church. The true gospel of the cross will remain an offence. There are signs that governments are objecting more to the cross. Wonder not if the world, including the professing church, hate you. Be concerned if all men think well of you. If it is trusted that God will end up victoriously, as mooted by some Psalms and by the prophets, then a defeatist view will be avoided. Is God going to vent his wrath on the whole of the strangers and then get all the citizens into heavenly territory after pulling up the drawbridge? It is obvious that evil involving the murder and oppression of innocent people will have to be redressed and that this is an argument in itself to support a need for judgment and the punishment of perpetrators. But is there also going to be a big breakthrough by kingdom evangelists so that it can be said that our God reigns? These matters will not be left entirely unresolved.

PART TWO

The Coming Kingdom

17 | Alternative Eschatology

Eschatology, the doctrine of last things, is a controversial subject. Whereas the bible is clear about history and the way of salvation, the manner in which the known world will end and what will happen afterwards is open to interpretation, and there are many. When unfulfilled prophecies are analysed, these make a lot of sense, but it is hard to piece these together to make a homogeneous whole. There are numerous scholarly commentaries in which a variety of views are expressed. For example, some see the Book of Revelation as having been fulfilled in the then contemporary Roman Empire (e.g. Nero was The Beast). Others place the same events in the distant future. Then some agree with both views, thinking that events will be replicated. So it might be wondered why it is necessary to write yet further on the subject as nobody can be certain of the exact chronology of the changing nature of the kingdom, and the juxtaposition of earthly, celestial, and resurrection bodies. In fact Julia Wood (now Mrs. Rees) found that conflicting opinions of preachers led to confusion. She said this left her with no clear understanding of the details of likely eschatological events. She was perfectly content, however, to leave these entirely in God's hands, knowing that the outcome would be much better than she could imagine. Perhaps this sums the subject up and disposes of a need to have a Part Two at all?

Well, hardly. First of all the subject is a corollary of Part One. And, as with a saved person without an awareness of a present invisible earthly kingdom, to review such a kingdom without considering its likely outcome is tantamount to standing on one leg. It will be essential to admit ignorance about some questions instead of staying stonily silent; not to claim to be right about everything especially when views are speculative. The aim will be to build an edifice that should leave readers with what is thought to be a simpler task than having to start from scratch themselves. They can dismantle and replace components if this is deemed necessary. The inspiration of scripture is recognised and a need to avoid selectivity accepted. Weight is placed on the words of Jesus especially in The Olivet Discourse. There is a slight problem with this in that as his words have been relayed by others, the past and distant future seem to have been telescoped but this can be disentangled. It is helpful that Jesus made it abundantly clear that the prophets are to be believed. As good examples, He started his ministry by lining up Isaiah's words with his missionary objectives, and after rising from the dead upbraided two disciples on the Emmaus Road for being slow of heart to believe all the prophets had foretold about Him.

The authenticity of the entire texts is taken as germane, unlike Luther's rejection of the Book of Revelation, and commentators who pick and choose what is and what is not reliable to fit in with their preconceived notions. Fundamental subjects include: a great tribulation; a great apostasy, a second coming of Jesus; judgment; a new heaven and a new earth; and a final consummation. Some contentious matters are: the nature of a rapture, two resurrections, and a millennium. An overview can prove helpful because by reference to it readers can assimilate whole chains of events without getting bogged down with complexities. As an example many find accounts of men on red and white horses; the blowing of trumpets; scorpion-type locusts; and speaking eagles, rather daunting and even incomprehensible. It is possible, however, to stand back and see that God is going to get angry with an increasingly evil world and, in consequence, some very nasty things will befall it. It fact, terrible and unimaginable things! It will not matter very much whether these are a result of self-inflicted damage to the environment or precipitate intervention by the creator. The former possibility can be anticipated from the apathy of G8 summits, and the spread of nuclear know-how to less stable countries. The latter is out of human hands but the portents of increasing sexual immorality, proliferation of HIV positive cases, increasing secularism, together with marked oppression and exploitation of the poor are all snubs to God's desire for peace and righteousness. Conversely scriptures can be drawn together indicating degrees of repentance by the nations in the future. Two examples of unacceptable teaching are a belief in British Israel and a modern form of restoration doctrine. In the nineteenth century and to a lesser extent the twentieth, there was a belief that the British were one of the lost tribes of Israel. In the Victorian era at the height of empire this view was rampant in the army amongst senior officers as it went hand in hand with colonial triumphal attitudes. As the empire and the state church faded in the twentieth century so did the popularity of this belief. However, it gained new ground amongst people in the new Pentecostal denominations and was one of the causes of a split in the Elim movement. Old Testament Jewish tribes are still identifiable and those who had Abraham's faith will be so raised. There can be no point in perpetuating an imaginary existence of "lost" tribes since all are one in Christ who receive him.

A current maverick organisation promotes outlandish views. These are ego-centric and part of a subtle self-esteem teaching that feeds the sinful nature. New super apostles, it is alleged, have emerged who will defeat the powers of evil and prepare a perfect kingdom into which Jesus will be invited to return as king. One prominent proselyte states that no prophet or apostle who ever lived equalled the power of these individuals . . . not even Elijah, or Peter or Paul enjoyed the power that is going to rest on many. To substantiate this development new "prophecies" are given and some claim that God spoke audibly to these individuals or even claimed that Jesus had appeared to them. Thousands of gullible people flock to their meetings. These meetings started in 1989 and the movement has achieved little more than drawing people into fantasies and producing a lot of hot air. The bible warns about the error of these kinds of activity in both Testaments. There is nothing new in this kind of approach. There was Bar-Jesus in Acts. Peter wrote, "But there were false prophets among the people, just as there will be false teachers among you . . . many will follow them in their shameful ways and will bring the way of truth into dispute. In their greed these teachers will exploit you with stories they have made up." This describes an early church problem and also fits contemporary behaviour of so-called evangelists in America, Britain and Germany. John wrote, "Many false prophets have gone into the world." These people are particularly keen to hold mass meetings in African countries.

With apologies to those who are not computer literate, it is proposed to tidy up our cluttered array of eschatological predictions. Some can be binned as not being supportable by the weight of scripture and depending on selecting verses that fit suppositions while ignoring others. Then the merits of the others can be compared, assessed or queried. It is possible to make much more sense of The Book of Revelation when our own mind sets are abandoned. For instance, it is too easy to assess future possibilities on the basis of a present status quo, as John Owen did in his day as relayed earlier in chapter 10. There is a need to allow that the end of all things will bring a much greater victory for God and his Christ than seems likely by the sight of our eyes in 2008. As a preliminary clearing up the following can be safely put in the bin:

Bin 1: British Israel as an aberration.

Bin 2: Modern Restoration Extremism as egotistic.

Bin 3: A form of Pre-millenarianism that supports a reign of 1,000 years in which the Jewish nation is restored, the temple rebuilt and Judaic sacrifices re-instituted. The earthly Jewish nation had its special relationship with God annulled when it rejected its Messiah. The Temple was left desolate in AD 70, and sacrifices ceased. There is now an Israel of God consisting of those Jews and Gentiles who are of the faith of Abraham. They are not all Israel who call themselves Israel. If these extremists were right it would mean a very retrogressive state of affairs. In a new and better covenant Jesus is not a High Priest of the old former Aaronic order but of the order of Melchizedek who offered bread and wine. There can be no going back to animal sacrifices. However it would be a mistake to abandon all thoughts of a literal millennium on the grounds of this dispensational error alone.

Bin 4: Post-millenarianism. This claims mankind will so improve, morally, culturally, and spiritually that it will bring in its own millennium or even that this has already started or is in the making. Since this contradicts scriptures predicting an initial falling away and increased apostasy, and is completely at variance with the sight of our eyes, it is not worth pursuing. It is only latterly that there is likely to be repentance on the part of the nations as two millennia have gone.

There are two theories that command support from large numbers of people, namely: another form of A-millenarianism and Pre-millenarianism. The former is supported by a large section of the established churches. Because of the difficulty of setting out a coherent summary of eschatology any attempt to present an infallible solution has its weaknesses. What might seem to be the most acceptable will still have its imponderables. Whilst criticising other's views, one's own view must not contain flimsy explanations. One must take seriously those points that stand up to scripture regardless of the camp these come from. Scriptures that pinpoint actual events must not be ignored or passed off as referring to something else. Just because a view can be demolished in some or many respects it does not mean it cannot contain a better explanation as to other aspects than what is claimed, wrongly, to be the absolutely overall correct solution.

Many academic theologians tend to overlook that their subject is concerned with the supernatural. As examples, the visions of the Apostle John recorded in Revelation are seen as being constructed by his own reasoning though they were conveyed to him from divine sources. John was the closest to the Lord of all the apostles and an obvious divine choice for the revelation of future events by the Lord Himself. Therefore, John was not consciously taking up and repeating something previously said by say Ezra, or Jeremiah. It was that the Holy Spirit can reveal the same truth independently to two or many. The fact that the precious stones on the High priest's breastplate and those in the buildings of the New Jerusalem are recorded in the exact reverse order to those connected by astrologers to the zodiac is not a thought up device by John. It should have occurred that it was the devil, who later reversed God's order in his world of occult counterfeit.

We must also be prepared to interpret scriptures from the viewpoint of a most wonderful God rather than from a propensity on our part to see that the worst will befall strangers. Is the redemption of a relatively small number of people all that Jesus will achieve? Will only a minor inroad be made against the Kingdoms of this world, who will then experience the wrath of God or will the title King of Kings mean that the nations are going to be won in the future? These are the issues that must be faced. So much so that the Book of Revelation merits a chapter of its own.

Bin 5: That part of the argument of pre-millenarians that rejects the existence of a present earthly invisible kingdom must be binned on the weight of scripture and of the arguments put forward in Part One of this book. Some of their arguments, however, are worthy of attention. So it is offensive, as some do, to describe all of their views as "rottenness". The A-millenarian view of a present kingdom lines up with the weight of Scripture but when supporters explain eschatology in a way that will dispense with any consideration of other views, this is unworthy. As examples, attributing the scripture about the feet of Jesus standing on the Mount of Olives as fulfilled when he walked on the mount takes a lot of swallowing. Zechariah was writing about the day of the Lord when all the nations will gather against Jerusalem, "and his feet shall stand in that day upon the Mount of Olives, which is before Jerusalem in the east, and the Mount of Olives shall cleave in the midst thereof . . . .half of the mountain shall remove toward the north . . . .the Lord shall be king over all the earth in that day." The Lord is King over his kingdom but presently Satan is the prince of this world that is of the powers of darkness that are yet to be overcome on the earth.

An argument that the first resurrection is when people are born again because they are described as having passed out of death into life seems plausible when looking at single scriptures, but does it stand up with others? Why, for instance, if this was the case, would Paul be unsure whether he had attained to the first resurrection when he taught a doctrine of justification by faith and of assurance? He also taught that citizens were still in Adam and that God's work would not finally be finished in them until the redemption of their mortal bodies takes place, presumably at resurrection? They have the first instalment: though outwardly wasting away, yet inwardly being renewed from day to day. For our light and momentary troubles are achieving for us a heavenly glory that far outweighs them all. Later he was sure he had attained. It seems he was referring to a crown or some kind of a reward. Therefore, a view that there is a separate future first resurrection should not be rubbished. So it is necessary to face up to two main different views about the end of the world. One being that we are already in a millennium even though that would make it a treble millennium. That the earth is destroyed. That Christ will come to judge the living and the dead, and that his presence will herald an everlasting eternal kingdom. The other is that there will be a first resurrection followed by an earthly millennial reign of Christ with his saints over the nations. It is only after the close of this time that Jesus will hand over the kingdom to his Father as it is subsumed into an everlasting heavenly abode. There are other questions to be dealt with such as who the nations are. Who takes part in the first resurrection, and why it will be necessary to cast the devil into the abyss at the beginning of the millennium when the beast and the false prophet are consigned to the lake of fire. Why is he then let out for a last fling?

These are all matters to be discussed in the following chapters. One can make justified criticisms of different schools of thought but not of every part of every thought. It comes to mind that Jesus said, "Why do you look at the speck of sawdust in your brother's eye and pay no attention to the plank in your own eye? How can you say to your brother let me take the speck out of your eye when all the time there is a plank in your own eye? You hypocrites. First take the plank out of your own eye, and then you will be able to see better to remove the spec from your brother's eye." So long as time is taken up in squabbling and claiming supremacy then the urgent work of the kingdom in preaching the true gospel to the inhabitants of the earth will suffer. It is not difficult in the light of this to understand the economy of Jesus about eschatology and his sense of urgency about the need to preach, that "the Kingdom of God is at hand."

It is intended to develop a round view of all these events in the following chapters. This will coincide in places with some views of the two credible main schools of thought but will be different altogether in others. Rather boldly, a possible chronology is attempted. Conclusions have followed a rationale that our great God and his anointed Christ will achieve a maximum victory over the powers of darkness and bring in times of perpetual righteousness. They will reverse and rectify all situations that have been brought about by interim victories of Satan over Adamic creation. These premises are supported by the word of God.

18 | End Days I: The Olivet Discourse

"Only in his home town and in his own house is a prophet without honour." *Matthew 13:57*

In his discourse on the Mount of Olives, Jesus spoke about the end days. These are deemed to have started at his first advent, then to continue up until the present and then extend into the future. There was no exact answer to the, "how long?" from the souls under the altar, but he made it quite clear that all that was to happen would stretch over a very long period. The discourse as recorded in Matthew 24, Mark 13, and Luke 21 concentrated on disastrous events and evil actions of men. This has led many expositors to think that all was, still is, and will be doom and gloom. But is all their pessimism justified? Before giving an answer much more needs to be made of the remarkable prophecy that provoked the disciples' question, "Tell us what will happen and what will be the sign of your coming and of the end of the age." As Jesus was leaving the temple he predicted of it that not one stone would be left upon another. It was a startling thing to predict destruction but to foresee such an unlikely and extraordinary finality was astonishing. It all came to pass within thirty-seven years just as prophesied. After Titus captured Jerusalem in AD 70 and massacred its population the Roman soldiers were idle. To give them something to do they were ordered "not to leave a stone upon a stone". If the Jews had believed Jesus this would have been averted. His followers did and, in consequence, escaped to Pella before the killing began. What befell the rest was not due to direct divine intervention but on account of an unwise and vicious revolt against the Roman occupiers.

Jesus went on to tell his followers that they would be flogged, killed and thrown out of the synagogues by the Jews and this all came about in a most terrible manner. This generation had falsely accused Jesus and demanded his crucifixion and just as its fathers had killed the prophets, so it killed the followers of Jesus. It is little wonder that their house was left to them desolate. Conversely, thousands of Jews were converted: three thousand alone just after Pentecost. No blame should attach to subsequent generations unless these repeat the offences. Unfortunately, there have been unjust persecutions of Jews ever since. This is because Gentiles have been equally evil. In the Olivet Discourse, there were remarks that referred to unfolding events far beyond the AD 70 calamity. False Christs, wars and rumours of wars, nation rising against nation, famines and earthquakes in various places would (and have) stretched over centuries. But the end was not yet. Citizens of the kingdom would be persecuted, put to death, and be hated on a large scale. Many would turn away from the faith and begin hating one another. False prophets would appear and deceive many. There would be an increase in wickedness. Citizens were told to stand firm. Then when the gospel is preached to the whole world there will be a testing of the nations and then the end will come. Jesus described a very terrible but short time during which there would be disasters the like of which had never been seen before. It would have to be short or nobody could survive. Then The Son of Man would come in the sky in complete visibility with great power and glory and all the nations of the earth would mourn. Angels would gather citizens from the four corners of the earth. The sun would be darkened and the moon would not give its light. Stars would fall from the sky and the planets would be shaken. When these things are over the fallen heaven and earth will pass away. Because the time will be short these will encompass no more than one generation.

Warnings do not seem to sink in. There have been periods in time when people enjoyed halcyon days and then, suddenly these were gone for ever. The Austria-Hungary Empire was like this in 1914. Then came war and allegorically the lights went out all over Europe. The empire and people's way of life was no more. Millions died. The cocooned life of colonial expatriates in South East Asia suffered a similar eclipse in 1941 when the Japanese invaded without warning. It will be much worse just prior to the end of the present world. The nations will be behaving just as they always had. In 2008 there are some signs that things are getting worse as far as, generally, sinfulness is concerned. The final world empire will conform to the feet of Daniel's image with feet and toes partly of clay and partly of iron. It will be divided on account of economic greed; religious and ideological differences; and mutual fear and distrust. Iron is strong and clay is weak and so a mixture akin to this will be in evidence in governments, powers, and peoples. This fourth kingdom will be punished and destroyed at the visible emergence of a divine kingdom of which Christ is king. What angers God is defiance of his authority; abuse of his creation; persecution and martyrdom of his own; and the worship of other gods. Motivation will either be satanic or egotistical. It is likely that altruism will still exist at public level in parts of the world such as in social services the treatment of delinquents, and by research the advancement of medical and general science. These factors might come to be regarded as weak by those with the spirit of Anti-Christ in due time: that is by followers of the Beast. This spirit was in the world in apostolic times and has reared its head over the centuries. Hitler destroyed the mentally defective and severely disabled, those who opposed him, and six million Jews. The Holocaust must be buried in prophecy somewhere because ten times more Jews were destroyed than were delivered from Egypt in the Exodus. Was it the time of Jacob's trouble? Then there have been huge acts of genocide already. A weak United Nations is unable to deal with tyranny because of disunity, corruption, misuse of resources, and absence of backbone. This means that millions of poor people suffer oppression while "constructive dialogue" is pursued with tyrants and war lords to little effect. In other instances, in Sierra Leone and Iraq, nations with strong armaments act unilaterally with the best intentions and depose despotic regimes or insurgencies. If these are beset with occupational difficulties, such as when populations are determined to kill each other, those on the sidelines become hypercritical.

God still works in situations such as the ending of apartheid in South Africa and achievement of peace after 75 years in Northern Ireland. Added to these things there is an ever increasing abuse of the environment brought about by industrial pollution as world populations increase and aspirations of greater personal wealth are encouraged. The increased carbon effect of air travel, internal combustion engines, and oil, gas and hard fuel emissions from power stations are damaging ozone layers. The resultant climate change is towards overheating and ice is receding at the poles. It is thought that the world is leaning over slightly more on its axis. Nanotechnology is pointing towards abominations of nature. Paedophilia is increasing as are crimes of a sexual or drug addiction nature. Euthanasia is being legalised and numerous abortions performed as matters of social convenience rather than medical necessity. Freedom of speech is under threat, in countries that were once proud of it, on the pretext of political correctness. On the plus side there has been the emancipation over the years of peoples from slavery and the improvement in the lot of some disadvantaged through racism. But the latter has not been without inverted reactions.

Jesus also touched on the rapture of citizens when they would come from graves and be taken into the sky and others left. It was left to Paul to elaborate on this event. This economy of Jesus was no doubt due to his priority to further the evangelistic work of the kingdom and to warn the nations about the prospective wrath that was in store for them. His pre-eminent desire was that the gospel should be preached and that citizens should be given a sense of urgency in carrying this out. His warning to the daughters of Jerusalem to weep for themselves can be joined with the discourse predictions about the city being surrounded by armies. The fact that Jesus himself, in his role as servant, said he did not know the time of the end, and that he predicted events that would take centuries to unfold, is proof that when he saw some standing with him "that would not taste death before the Son of Man comes in his kingdom," he must have been referring to a contemporary event. This statement, in any case was not part of the discourse. It was made much earlier. Immediately after this Peter, James and John saw Jesus transfigured on a mountain. However, it was more likely that Jesus was referring to a post-resurrection appearance. The invisible spiritual kingdom had just come into history. The first thing he spoke of to his followers was the Kingdom of God. Prior to the Olivet Discourse, Jesus had pronounced seven woes upon the hypocritical Pharisees and Sadducees. It is not that their deficiencies should be harped upon at this distance of time except to provide a salutary lesson.

There are similarities to the subsequent behaviour of the professing Christian Church. This was raised in Part One. The accusations are that they did not practice what they preached; they put burdens on people's shoulders; they sought to be seen by showing off and displaying titles. In their case it was "Rabbi" and "Teacher" and "Father." Titles used by the visible church now include, "His Holiness," "Your Eminence," "His Grace," "Most Reverend," and again, "Father." They shut the kingdom in men's faces, they don't go in themselves, or let others go in. This is reminiscent of nominal religionists who substitute the true gospel of the kingdom by a gospel that is not a gospel. Gospels of works, of ritualism, and ostentatious ceremony are not gospels at all and hide the true kingdom. Jesus said they make converts who become twice as bad as they are!

Still referring to the immediate Jewish generation, Jesus said that it would kill, crucify and flog his followers and cast others out of their synagogues, and pursue them from town to town. Paul was taking part in this persecution when he was halted on the Damascus Road. This all came to pass and it is worthy of special note that these vicious acts continued for thirty seven years until the destruction of Jerusalem in AD 70. It was only when the Romans realised that early Christianity was not just a Jewish sect that it removed its protection from them. Jews were the only people exempted from worshipping Roman gods. Then Nero began a diabolical and atrocious martyrdom of true believers resulting in the deaths of Peter and Paul. Romans were amongst the cruellest people who ever lived. Nero was considered by some to be the Beast of Revelation, described as "who was and is not." There was a belief he would come back to life again. Without taking up a comparison with the beast but staying with a temporary concept it is interesting to note how tyrannies come and go. A few examples are The Spanish Inquisition, the reigns of Queen Mary I, and Queen Elizabeth I, daughters of King Henry VIII successively Catholic and Protestant. A whole system can terrorise for a period and then vanish as was the more recent case of German National Socialism (Nazism). No blame can attach to subsequent generations unless these commit the same crimes. This was the case in the time of the "woes". whose forebears had killed the prophets. It is not now denied by open-minded Catholics that the medieval church failed abysmally over the centuries. In order not to be seen as exaggerating or being unnecessarily partisan it is better to publicise self-criticism. Cardinal Newman was frank. "I might have said that the Middle Ages were as virtuous as believing. I might have denied that there was any violence, any superstition, any immorality, any blasphemy during them. And as to the state of countries which have long had the light of Catholic faith, and have degenerated. I might have admitted nothing against them, and explained away everything which plausibly told to their disadvantage. I did nothing of the kind!" So it will be seen that evil comes and goes within systems. However, it is interesting that as doctrine this church added a whole collection based on tradition. It has only made minor adjustments by stealth and so contributes to hiding the true kingdom.

It is not alone because Protestantism was also guilty of similar degeneracy but remains just as self-justified about it. One reviewer criticised a balanced script that laid blame for persecution equally. He wrote that whereas Catholics deserved to be put to death, the Protestants similarly treated were martyrs! Elizabeth of York refused to renounce her Roman faith and was gradually crushed to death by having layers of large stone slabs placed over her body. More recently a Protestant commentator was more balanced when writing in a well circulated denominational journal A minister told him that any man who kicks the Pope over the Derry wall was welcome in his pulpit. The writer went on to state Protestants who fail to live up to the Christian life engage in things that are negative. (presumably he meant unprofitable criticism). They have an axe to grind and seek a platform to do so. They are seen by other Christians as cranks, odd balls and fanatics, (contributor to *Our Inheritance Ministries,* 2007). He is as frank in his writing as John Newman. However, neither man would be prepared to give an inch in what he believes is the only truth. The fact is that both have some of it. One has added to it and the other taken away from it. Neither should be seen as opposed to God. Their denominations became polarised and more concerned to develop what they thought to be a watertight system. The trouble has been that they have not submitted to the kingdom sufficiently by prioritising that God's will should be done on earth as it is in heaven.

Again this week Pope Benedict XVI declared that his is the only true Christian religion despite all the efforts of Vatican II to promote ecumenism. The small print always showed that this had to lead to the universal acceptance of the Pope as the head of all Christians. The authority for this is based on what is labelled The Apostolic Succession, a premise that only somebody in the "Holy Orders" of Catholicism can administer "sacraments" because ordination is only by the laying on of hands in a continuous process from the apostles to the present time. Moreover the Pope is the direct successor of St. Peter. What is that church's justification for such an exclusive and privileged assertion? According to Archbishop Sheehan D.D., in his *Catholic Apologetics*, there was no discoverable proof in early centuries "because the church was in its first fervour" and therefore was too pre-occupied. It was only in the fifth century at the Council of Ephesus that one Philip, the Legate of Pope Celestine said, and no voice was raised in protest: "no one doubts but nay, all ages know, that the holy and most blessed Peter, prince and head of the apostles, the pillar of faith and the foundation of the Church, received from Our Lord Jesus Christ the keys of the kingdom . . . his successor in order, and the holder of his place our holy and most blessed Pope, Celestine has sent me." Readers are asked to take a jurist's attitude to this as suggested at the outset and decide whether this is sound evidence upon which to brigade one church as the only true one and to denigrate all of the others. Furthermore, to judge whether this continuing intransigence is harming the work of the true kingdom. The starting point ought to have been that there should be a worship of the one true God and an ardent furtherance of his present kingdom in its fight against the powers of darkness of this world. This will be the outcome at the end of time whether this kind of squabbling continues or not.

This will become more obvious as the Book of Revelation is discussed in the following chapter. In the Olivet Discourse Jesus also said that Jerusalem would be trampled on until the times of the gentiles were fulfilled. Although some hopes were raised when General Allenby freed Jerusalem from the Ottoman Empire in 1918, this only resulted in a British Protectorate. Some expositors made a similar mistake to that made by John Owen about the Evangelists. They considered that prophesies concerning the return of the Jews to their own land in this world were not likely to be fulfilled. This was placing their own reason above trust in God's word because against all expectation the Jews went back to their own land in 1947 and the times of the gentiles was over! Some remarks of Jesus hint at a future earthly millennial reign although stronger references can only be found in the psalms, the prophets and The Book of Revelation. Jesus said that the meek would inherit the earth. He said that at the renewal of all things when the Son of Man sits on his throne those who had followed him would take part in the judgment of the twelve tribes. His promise that everyone who had left houses or brothers, or sisters, or father, or mother, or children, or fields for his sake would receive a hundred times as much and would inherit eternal life must refer to a future period of renewal and not to mortal existence.

Jesus also mentioned that the Father had conferred a kingdom upon him. He also told the Jews that a time would come when they would see the angels of God descending and ascending on the Son of Man. All these sayings of Jesus must be mulled over deeply in the manner that Mary kept in her heart the predictions of Simeon and Anna. Jesus also referred to a great tribulation, apostasy, martyrdoms, judgment, a rapture, a second coming; rewards, and world-wide dissemination of the gospel. All these things will be returned to but two are taken further here. That two will be in a field and one taken and one left ties in with Paul's statement that at the end believers will be caught up to meet the Lord in the air. They will be changed in a moment, in a twinkle of the eye. It also fits in with the angels who will gather the elect from the four corners of the earth and corresponds with the statements about the first fruits and the grain harvest. This will be a time of judgment, but what will be the nature of this? At law persons are indicted for an offence. At the time the Roman Principate was in its infancy, being founded in 27 BC. Under this proconsuls like Julius Caesar held an *imperium* that gave him sweeping powers of jurisdiction. Nevertheless there had to be accusers and shortly after the Discourse Jesus would have been accused by not more than three or four members of the Sanhedrin. In eternal circles Satan was an accuser but Jesus is now the judge. It seems clear that people must be judged immediately upon death. It is appointed unto man once to die and after this the judgment. Otherwise there would be no means of knowing where to direct souls for the period between death and eventual bodily resurrection. For those who go into God's presence there can be no question of any subsequent punishment. As an obiter dictum: this will be as a soul and not as a member of any faction. So for these individuals final judgment can only mean some sort of assessment for reward or admonition.

It is purely speculative if a suggestion is made that there might be rewards for effective and properly motivated work in the kingdom and that others might have what they never had taken away from them. That would be responsibilities. Their unprofitable activities would be burned up like hay, wood, and stubble. As to how others are to be judged is a moot point. Firstly there has to be dire punishment for the vile, evil, and wicked. The recent controversy involving a British Christian national event, caused by a teaching that God will punish nobody, is harmful. Emphatically, the judgment and destiny of the wicked is in the hands of a holy God who has the unpleasant task of exacting vengeance on behalf of his people and particularly his martyrs. He must also punish those who deny his sovereignty and follow after other gods. This is not really an appropriate subject for human debate. An esoteric point is to wonder what happens to those who never heard the gospel, or having heard it, ignored it without positively rejecting it. After this some would have led helpful lives. It would be pure speculation to imagine there might be some leniency or clemency exercised. It would be the prerogative of God alone to allow some to escape the eventual lake of fire. It is not necessary to doubt that some hard-line individuals would be disappointed if this proved to be the case!

Immediately following the Discourse Jesus gave the parables of the ten virgins, the talents, and the sheep and the goats. The latter is relevant to what has already been discussed except that Jesus was not explicit about who he was referring to when he spoke of those who gave cups of water, clothing, cared for the sick, and visited those in prison. Could it have included those who were ostensibly non-believers? The two first parables are severe warnings to servants who do not keep watch, who beat the other servants or backslide into bad ways. Again are these people genuine believers? Hardly, because Jesus speaks about them being beaten with few or many rods and even cut to pieces. If these are true believers the words seem both incompatible with redemption scriptures and out of character with the love of Jesus. There is not an obvious answer to this. As to the manner of the second coming, Jesus said he would come in the clouds with great glory and this is consistent with the words of two angels after his ascension, "Why do you stand here looking into the sky? This same Jesus, who has been taken from you into heaven, will come back in the same way you have see him go into heaven". Jesus made no reference to a one thousand year millennium or to what degree of success there would be in seeking to win over the nations. It might be questioned why Jesus did not comment on whether the nations would be won in a substantial manner or not. There are several answers to this. The most obvious one is that it is not up to anyone to pre-empt the choice of people. As we shall see later, some will already have acknowledged God's sovereignty. Others will oppose it for a period because of unbelief or allegiance to the Beast. Finally, all will become aware of the existence of a one true all-powerful God, but will re-act in one of two ways. Either they will repent and transfer their allegiance or they will continue to curse God regardless. God is aware by foreknowledge how people will re-act and the extent to which the inhabitants of the earth will repent and worship him, but understandably such knowledge is retained in the secret of his own counsels. We do have encouraging indications that conversion of the nations will be substantial, and I shall return to this subject in the following chapter.

Before doing so some of the finer points of the Discourse will be separated as far as possible by attributing words either to the destruction of Jerusalem in AD 70, to interim history, or to the end times. Matthew Chapter 24 will be used and instead of the practice of showing references in the index, scriptures will be quoted in the text. The "not a stone upon a stone" prediction in verses 1 and 2 was fulfilled in AD 70. It is obvious that verses 4 to 8 refer to ongoing history. Verses 9 to 14 refer to the great tribulation that is to befall the world. Verse 14 states that the gospel is to be preached to the whole world as a testimony to all the nations, as a precursor to the end, and links in with Revelation. This must include the efforts of the missionaries over all the centuries but it seems that as persecution increases that God will escalate witness by divine means. Firstly there will be the two witnesses. After being killed by the Beast they will be brought to life again (Rev. chapter 11). Terrified nations will then glorify God (Rev. 11:13). Then the first of three angels (Rev. 14:6) will take the eternal gospel to all those living on the earth – to every nation, tribe, language and people. Verse 15 must refer to AD 70 as there is no longer a temple. The abomination of desolation referred to by Daniel, which Jesus had in mind, most likely fits in with pollution by Antiochus Apiphanes. In Jewish thought pollution is the cause of destruction. Manasseh's pollution led to the destruction of the first temple (2 Kings 21:7–15). Verses 16 to 20 seem to be a warning of the approaching Roman armies. There was a previous siege by Cestine Gallus in AD 66. Verses 21 to 25 must refer to a future great tribulation (see Rev. 7:14 and Rev. chapter 13). Then verses 26 to 29 speak of God's vengeance on the Beast and disobedient nations (Rev. chapters 8 and 9). Verses 30–31 relate to the second coming and the rapture of citizens (Rev. 20:6).

There are so many chapters and verses in the Book of Isaiah that complement New Testament predictions about the end times that these cannot all be referred to but some of them are extraordinarily compatible. A short example is as follows: "See, the day of the Lord is coming – a cruel day with wrath and fierce anger – to make the land desolate and destroy the sinners within it. The stars of heaven and their constellations will not show their light. The rising sun will be darkened and the moon will not give its light" (Isaiah 9:13, 10). The sign of the Son of Man appearing in heaven was foreshadowed by Daniel (Daniel 7:13) and more positively (Daniel 2:44, 45). The prophets had no doubt that there was to be an earthly kingdom. Jesus spoke with all that authority which he alone could possess and he was confident that all these things would be fulfilled.

19 | End Days II: Revelation of John the Apostle

"For I dipt into the future, far as human eye could see,
Saw the vision of the world, and all the wonder that would be."
Locksley Hall, Alfred Lord Tennyson

It is clear that Revelation is based on a series of visions experienced by the Apostle John. The risen Lord Jesus dictated letters to him for seven churches and conveyed other details of the future through a series of angelic agents. These experiences appear more authentic and authoritative than if these were regarded as just a scholastic exercise influenced partly by Old Testament writings. Any connection with the latter would, therefore, simply amount to consistency of inspiration by the Holy Spirit. An attempt is being made in this chapter to interpret this deep and challenging book by placing predicted future events in a sensible chronological order. It is not surprising that some ordinary readers, when reading straight through Revelation, retain salient points in their minds but are left without appreciation of their coherence. For this reason it is thought there is merit in re-jigging the order of the visions in an obvious and logical manner. First of all there are several chapters that do not lend themselves readily to identification in the final chain of events: these are:

Chapters	Events
1, 2, & 3	Christ's letters to the seven churches
4 & 7	Visions of Heaven
10, 12, & 17	The Little Scroll
	The Woman on a Horse
	The Woman on a Dragon

The extent to which the letters to the churches by the risen Lord impacted upon the actual churches to which these were addressed in John's day cannot be traced from historic records. Their content would not have appeared in the canon of scripture unless it was intended to apply to types of church throughout time as well. Some excerpts seem to apply to particular periods while others are obviously general and universal. What stands out is there never was nor ever will be a perfect church. Jesus saw that churches would always contain a mixture of true and false people, and this confirms that the terms "church" and "kingdom" are not synonymous. This also lines up with the parable of the dragnet. There should be repentance or the Holy Spirit would be withdrawn. There will always be a hard core of faithful ones but even some of these will have weaknesses. The weak will be too tolerant of pretenders such as Jezebel who says she is a prophetess and is not. Some are lukewarm. Some are neither hot nor cold. Some have forsaken their first love. On the positive side some are hardworking and persevering. They are loving and faithful. They keep God's word even though they have little strength. There would be imprisonments and persecutions. In reality this has been ongoing. There would be an open door to preach the gospel which no man could shut. One promise related to the end times which was, "I am coming soon." Jesus referred to "The Nicolaitans" and to "overcomers", the former being oppressors (devourers) of others within the churches. They were antinomian in character and described at Ephesus as grievous wolves who did not spare the flock and drew disciples after them. Their activities were a feature at both Galatia and Corinth. There was greed for the wages of unrighteousness. There were liars and those who were determined to capture the ministry. The seven apostasies in The Book of Judges confirm that wickedness amongst God's people has always been an endemic fact. It is easy to identify both Nicolaitans and Balaamites on today's scene. As to overcomers, there are many strange notions about who these are, ending in unwelcome "them and us" situations. This description must refer to all true believers because otherwise the others would end up in the lake of fire: a most unscriptural outcome. The title occurs in each of the seven letters and seven times in John's gospel. It is also to be found five times in Romans and six times elsewhere in Revelation: so it is important. The reward is eternal glory. There are references to a crown of life in the letters. This is a reward that, unlike salvation, can be earned by those who are faithful unto death and can also be lost. There is eschatological significance in the title "overcomer" in that the letters to the churches deal with all ages including work in the present kingdom.

The rewards are: to the church at Ephesus the right to eat of the tree of life in the paradise of God. To the church at Pergamum, the gift of hidden manna and a white stone with a secret name on it. To the church at Thyatira, authority over the nations. To the church at Sardis, never to blot a name out of the book of life. To the church at Philadelphia, to have the name of God and of the new Jerusalem written on them. To the church at Laodicea, the right to sit on Christ's throne. And so it becomes abundantly clear that those who belong to Christ shall not come into condemnation because they have already passed out of death into life during their mortal lives. It can also be seen that at the Parousia it is rewards or lack of these that will be adjudicated rather then entering into any judgment of condemnation.

Several problems are encountered in trying to set out a chronology of the Book of Revelation and for this reason most writers avoid even making attempts to do this.

The following shows a possible sequence of events at the end times, according to Revelation:

Sequence	Chapter	Event
		Great Tribulation
1	3	Beast given power over earth for 42 months. Kills citizens
2	11	Two Witnesses killed and raised to life
		Rapture of faithful (see 8, 9, 10, & 11)
		Vengeance of God
3	5 & 6	Vision of Scrolls
4	8 & 9	Trumpets
5	16	Bowls of Wrath
6	18	Three Woes
7	19	Armageddon
		Victory of Christ and the Martyrs
8	19	Hallelujah
9	14a	Saved in Heaven
10	14b	Angel with Gospel
11	15	Song of Moses
12	19	Beast and False Prophet in Lake of Fire
13	20	Satan chained in abyss
		Second Coming of Christ
14	20	A restored earth.
15	20	1,000 year reign of Christ with faithful
		Consummation of Kingdom
16	20	'Satan released' – stirs nations
17	14	Harvest of the earth
18	20	Dead are judged
19	20	Satan and wicked thrown into lake of fire

| 20 | 21 | New Jerusalem |
| 20 | 22 | New heavens and new earth |

Notes:
(i) See the table below, which details seals, trumpets, and bowls.
(ii) It must be the Beast who initiates the killing; vengeance then follows.
(iii) The victory is not just that the martyrs overcome but that the nations are won.
(iv) People are assessed at death and judged at 18 above.

The redeemed reign with Christ, which is "the first resurrection", and the rest of the dead, are not raised until after the 1,000 years.

It can be seen from the following table that the seals are just a written record and that the trumpets are starting signals. It is only the outpouring of the bowls that involve positive action. The seals do not need to correlate but the trumpets and bowls can be seen to line up.

Correlation of Seals, Trumpets and Bowls

Item	7 Seals	7 Trumpets	7 Bowls
1st	White horse Conquering	Hail, fire, blood and one-third trees burnt	Painful sores on those with mark of beast
2nd	Men slay each other	Mountain hurled in sea one-third blood	Sea of blood
3rd	Shortages	Polluted water One-third die	Polluted water
4th	Famine, plague and sword	One-third sun, moon turns dark	Sun scorches people
5th	Martyrs under the altar	Locusts come from abyss sting wicked	Darkness, agony
6th	Earthquake, moon and stars shake	Angels kill one-third mankind	Evil spirits deceive
7th	Silence, earthquake	Earthquake Kingdom has come	Earthquake It is done

The following is deemed to be a possible explanation of the events set out in the two tables above. But first of all some reservations are discussed based on views of other commentators.

The time of the judgment is a conundrum. If it was placed in juxtaposition with the Parousia it would not coincide with the Harvest of the Earth and the sheep and goats episode. The just possess the Kingdom prepared for them. The wicked sentenced to the lake of fire. However, this poses a problem. The rest of the dead are not raised for 1,000 years. Only at this later point of time are the raised dead judged according to their works and checked to see whether their names are written in the Lamb's Book of Life. To persist with this order it must be assumed that the unsaved dead are accused but not actually sentenced. But then those who support Satan at the very end would not have been judged. The only way to circumvent this argument would be to regard the millennium as outside of the present worlds and, therefore, the rebellious armies would go automatically into destruction with Satan. The alternative view would be to place the judgment at the very end. This would mean that Jesus would not judge the world at once at his coming and that there would have to be a different process for rewarding the just by giving them responsibilities in the millennium. This is the preferred view taken. Another not very popular view is that only the martyrs and special overcomers take part in the first resurrection and that weak Christians are not raised for a further 1,000 years and are judged with the rest of the dead. This is not feasible because all "weak Christians" would already be in heaven at death with celestial bodies!

It would seem that as the world gets worse and worse as predicted in many scriptures it will come under the complete domination of a character called "The Beast". He will persecute the real citizens of the kingdom and put many to death who will not accept his mark. Following this escalation of evil and wickedness God decides to execute vengeance upon the perpetrators. It is suggested that believers will suffer the Beast's attacks because many are faithful unto death. This is the great trial that will beset them. It is evidence of the superiority of the new person as opposed to the weakness of Adam. True believers who elude the slaying will be specially protected and escape God's vengeance. During the Great Tribulation of the Beast the two witnesses will appear and be given great powers. It is favoured that these are to be Moses and Elijah. If it is not these personally then it must be with their spirit (as it was in the case of John the Baptist) because the supernatural abilities prove this. They will be able to bring fire from their mouths, to shut up the heavens so that it does not rain, turn the waters into blood, and bring down every plague upon the earth. These are the things Moses and Elijah did. When they are killed by the Beast and the world rejoices, they are brought to life again and taken visibly up to heaven. The people are terrified. Following an earthquake that kills some people the survivors glorify God.

There are a number of scriptures indicating that faithful believers will not have to endure the effects of God's vengeance and that is why they are at this point caught up to join the Lord in the air. As Jesus said, "There will be two in a field and one will be taken and the other left." Those who accept Christ after this and those still on earth but not yet saved will live through God's vengeance but enjoy some providential protection (e.g. from plagues) because those affected will be the ones with the mark of the Beast. Christ's enemies will be destroyed at Armageddon. It could be that they turn on one another. The Beast and The False Prophet go into the lake of fire and Satan is bound and put into the abyss. Immediately after this the earth will be scorched and destroyed, but only in the sense that the earth was said by Peter to be destroyed at the deluge. What will emerge is a restored earth.

The basis of this reasoning is that God will turn the tables completely on Satan. The scriptures are clear that creation is groaning for its own redemption. Prior to Satan's victory over Adam creation was perfect. All animals fed on grass and, therefore, the wolf laid down with the lamb. That situation will be returned to and then gives God a reversal to his original will. Those already dead in heaven will include those who came out of the Great Tribulation clothed in white robes. They are the first fruits. They will sing The Song of Moses, which is an indication of a similar deliverance to that of the children of Israel from Egypt. They suffered grievous persecution. Boy infants were ordered to be killed. When the nations round about saw the mighty acts of God especially at the Red Sea they turned to the true God. The Israelites went into the place of worship. So the song in heaven can be taken as rejoicing about the triumph of The Lamb and the martyrs, the destruction of the powers of evil, the conversion of the nations (one of the Lord's great objectives), and coming into the very sanctuary of heaven itself.

Then follows the millennial reign of Christ which will be dealt with fully in the following chapter. At the end of the millennium will come the harvest of the earth by those with the sickles. The grain harvest is reaped and this will comprise those whose names are written in The Lamb's book of life. Wheat is put into the garner after the chaff is burnt up. The grape harvest is a vintage that is trod out in the winepress and typifies the fate of the wicked. Then comes the judgment. The dead, both small and great, will stand before the great white throne. As already stated it is worthy of repetition), it is obvious that people are judged when they die. It is appointed unto man once to die and after this the judgment. Otherwise there would be no means of knowing where to send departing souls. The judgment must be one of sentencing or commuting. It makes no sense that souls once in heaven will be liable to punishment. The redeemed will have been given rewards and responsibilities or suffer loss of these (that which they do not have will be taken away from them). If they have only built hay, wood and stubble on salvation's foundation it will be burnt up but they will be saved as by fire. All the redeemed will have possessed the kingdom prepared for them at the coming of Christ and will have reigned with him.

At the end of the 1,000 years the sentences passed will be executed. All the dead will stand before the judge after they have been raised. Jesus will say to the goats, "Depart from me you cursed into everlasting fire that was prepared for the devil and his angels." It does not seem appropriate to dwell on the punishment of the wicked except to observe that they must deserve whatever a just God decides. Then comes the most mysterious problem. The saints will have reigned with Jesus as King on a restored earth and the rest of the dead will not be raised for 1,000 years. It seems there is a chink of light for those who fall short and have missed the first resurrection. Those with a concern for all people will appreciate that God is not willing that any should be lost. For example, Farrar argued, controversially, that the spirits in prison preached to by Jesus were those who died unrepentant before Christ commenced his earthly ministry and had not previously heard the gospel. So without preaching a doctrine of a second chance there is still a chink of light in the fact that it can be implied there will be some whose names are written in the book of life. Otherwise what would be the point of checking out those whose names were not so recorded? Are these people unfaithful believers? (hardly, because their spirits would be in heaven). Are they those who never heard the gospel and will be judged by works? Or are they the parents of some of us who never came to the Lord but were good parents and apparently honest people? These are things we cannot be sure about. In his wisdom the Lord Jesus gave total priority to urging his followers to preach the gospel throughout the world and to warning the nations of the consequences of refusing to hear them. These considerations are of prime importance and must be treated here as such. Once the divine urgency for the salvation of the nations and the defeat of darkness is dealt with other prospective events will fade, comparatively, into a mere satisfying of inquisitiveness.

A casual reader of Revelation could be forgiven for thinking it is just an indictment of wickedness and an account of how God's wrath will come upon evil doers. Furthermore, true believers will be slaughtered and then the world will terminate with the death of its inhabitants. In other words, all is doom and gloom. But this is far from the case! This deep and challenging work has been the victim of opinions and guess work on the part of interpreters over the centuries. Several of the points that have been made already need further clarification. Supporters of pre-millennialism have made mistakes in proclaiming a restored nation of Israel on the present earth and some even believe in the restoration of animal sacrifices. This leaves them wide open to criticism by A-millennial protagonists who do not believe in a 1,000 year reign.

The stance being taken here is that God will get the complete victory over Satan, hell and death. The Israel addressed is the Israel of God that is all those who have the faith of Abraham whether Jew or Gentile. Jesus knew that they were not all Israel who called themselves Israel. Another point is that the A-millennial supporters got it right when they said that there is currently an invisible spiritual kingdom on earth already. However, what is missed by both is that there is a difference between a present earth, a restored earth, and a new heavens and earth. The 1,000 year reign will be of the converted nations by Jesus as King and his resurrected saints. It will be post fallen earth and so spiritually restored.

When believers die they are clothed with a celestial body. However, the Lord has a resurrected body. If the saints were not going to return to earth what would be the point of resurrecting their mortal bodies and changing them in a moment if a celestial body was appropriate for earthly service? Then there is the retrograde argument. Why would they want to go back to earth when saints are already in heaven? God's ways are not our ways and we find it very hard to accommodate to these. John's account of the Lord's Supper is markedly different from that of the three Synoptic evangelists. They omit the feet washing and he omits the bread and the wine. Yet he was the one who leant on the Lord's breast. As one of the sons of Zebedee he had learnt a salutary lesson. His mother had asked Jesus whether her sons could sit on his right and left hand in the kingdom. Jesus said it was not up to him and in any case anyone who wanted to be great must become the servant of all just, as he had come to serve and give his life a ransom for many. It was Matthew who let the cat out of the bag about this incident. John recorded the fact that Jesus had to ram this point home to the rest of the Apostles by washing their feet at the Last Supper. The redeemed of the Lord will be totally motivated to serve the Lord in reigning over the nations as a strong preference to basking in heaven, as delightful as this must be.

Then the A-millennial camp argues that there is no point in letting Satan out to have a second bite. But have they thought deeply enough about God's total triumph and vindication? The vital reason why Satan is not thrown into the lake of fire with The Beast and the false Prophet is that there must be a final show down between regenerate mankind and the same old Satan. This time he will not succeed in deceiving the new nature. So who are those among the nations like the sand of the sea who fall to Satan's blandishments? These will be those whose submission to the rod of iron was partly due to terror at God's mighty acts and partly knowing which side their bread was buttered on. Unlike those who, realising there was indeed a one true mighty God, still decided to curse him, the cunning ones, still being only of a sinful nature, will go along with what suits them. There is a vast difference between the sorrow of the world that worketh death, and a godly sorrow that works a repentance not to be repented of. These two characteristics separate those who will join the armies of the Beast and those who will acknowledge Jesus as King.

Some further comments are necessary concerning the events described above. It is unfortunate that studies in the past have been hampered by those whose views have given occasion for criticism. For example, by falsely linking historic persons with John's apocalyptic characters. There was failure to distinguish the difference between portraying those in history with similar characteristics without being the actual characters described by John. They also failed to spot that all the other events that are predicted to occur around the Beast have so far been singularly absent. For instance, the Beast has been identified as Nero, Napoleon, and Hitler but these individuals did not have authority over the whole earth or demand that people worship them or be put to death for not worshipping them.

Although these historic characters were not those identified by John they were undoubtedly influenced in a demonic manner. This led them to label others as Anti-Christs quite falsely and demonise them unjustifiably in the sight of many. They were not martyrs but certainly victims who were not guilty as accused. A medieval instance is the treatment of Jews in Moorish Spain, once it had been reclaimed by Spaniards. The Jews had thriving communities under the Moors. What happened next has been swept under the carpet and it is only one tourist in a thousand who after being shown a historic synagogue in Cordoba will ask where the Jews are today. The answer is that there are not any. Hitler wrongly demonised the Jews and Stalin demonised the middle classes and the landed peasants. The Pilgrim Fathers fled to America because they saw themselves as advocates of religious freedom, and so they were. However it was not very long before some of them became little ogres in their own right. The history of the Massachusetts Bay Colony shows that it is possible to allow freedom of conscience when that means freedom to believe what authority wants you to. There was also the regrettable trial of "The Salem Witches". Nevertheless, America stands today as the country with the greatest freedom: champions of liberty of conscience. This is what has been lacking down the ages on the part of religionists. God has never shown any desire to force people to do anything. He is sad when they sin. Jesus once said that he had held out his hands all day long to his people: "but they would not." That is why the book of Revelation records the words. "Let him who does wrong continue to do wrong; let him who is vile continue to be vile; let him who does right continue to do right; and let him who is holy continue to be holy." As a man sows so shall he reap. The one who sows to please his sinful nature, from that nature he will reap destruction; the one who sows to please the Spirit will reap eternal life. An actual personified Beast is to emerge in the end days as has been shown.

The greatest nuggets in Revelation are those that speak of the repentance of the nations and the rejoicing of the martyrs about this event. That this rejoicing is joined to the Song of Moses is the most beautiful of all the chapters (15). The subject addressed here is The Book of Revelation but the study of this book must spill over into the following chapters. The total calendar has been set out here and the events up to Christ's coming covered in detail. However, Christ's reign on earth and the final consummation, although covered in Revelation, have only been touched upon. It has been decided that fuller comment including about the fall of Babylon can be more appropriately dealt with in the chapters to follow.

20 | The Fall of Babylon

"The angel of Death spread his wings on the blast,
And breathed in the face of the foe as they pass'd;
And the eyes of the sleepers waxed deadly and chill,
And their hearts but once heaved, and forever grew still!"
—
"The widows of Ashur are loud in their wail,
And the idols are broke in the temple of Baal;
And the might of the Gentile, unsmote by the sword.
Hath melted like snow in the glance of the Lord."
The Destruction of Sennacherib, Lord Byron

The Fall of Babylon provides a deep, deep, lesson for everyone. A vision of such a violent future catastrophe is something that, naturally, is recoiled from. It is of very much more consequence than the event itself. That is why it merits its own chapter. It will go towards bringing this study to a climactic conclusion. To begin with the identity of the biblical Babylon has been well hidden. A description of its fall appears in the eighteenth chapter of Revelation. But as to its identity, the early church thought the title was a disguised name for Rome, and more particularly for the Roman Empire. The reasons given for such a supposition were that it would have been too dangerous to mention 'Rome' openly at such a time; the Emperor Nero was beast-like; and the Jewish Talmud used such a device. However, it is absolutely clear that the oligarchy referred to was a future political and commercial monster that will oppress and cheat, and deceive, both its own people and the nations it dominates.

John's allusion to 'Babylon' is now known to relate to the future because the Roman Empire is defunct and no bowls of wrath of a kind described in Revelation were poured upon it during the period of its existence. Unsurprisingly, an argument that John referred to ancient Rome as 'Babylon' assumes he used his own reasoning but falls into the trap of wrongly supposing the source was his own scholarship instead of it being a direct message from the risen Lord Himself (a trap already warned of over linking the Beast to various historic personalities). There is a further identification error that subtly protects the real culprit. It has been alleged, widely in Protestantism, that Babylon is the Roman Catholic Church. This allegation was dramatised in Rev. Alexander Hislop's book *The Two Babylons* published in the nineteenth century. It was the result of a mammoth piece of research that found similarities between ancient Babylonish mystical religious practices and new objects of worship introduced into the Roman Church by Constantine. Whatever is made of these findings that occupy most of his book, the final and seventh chapter failed to link convincingly the Beast of Revelation with Catholicism. The woman on the Beast was claimed by some to be a harlot church whereas the bible asserts that the woman is the wicked city (the capital or hub) that sits on the rest of the evil empire.

There is no excuse for the medieval excesses of the Papacy and the respected Cardinal Newman, as mentioned elsewhere, did not attempt to make any. This was not Babylon, and anyway there were atrocities committed by other denominations. Consider the barbaric treatment of Edmund Campion, the Jesuit, hanged, drawn and quartered at Tyburn, and the crushing to death of Elizabeth of York by the application of progressive layers of paving slabs. All she was guilty of was refusing to renounce the Catholic faith she had been brought up in. Then there was John Calvin's vicious treatment of an Arminian who had the misfortune to visit Geneva. He not only had him burnt at the stake but asked that the wood should be green so that the process would take longer!

When the actual Babylon emerges it will be a trading monopoly controlled by the Beast. There will, at the same time, be some form of religion in Babylon. Who knows what it will be? It could be a degenerate Papacy or possibly a psuedo-charismatic circus: a bit like certain current false prophets only much worse. What seems to be missed is that whatever it is, the Beast will turn on it and destroy it in the manner that Hitler destroyed all secret societies in Germany. So this hypothetical religious body cannot be described as 'Babylon'. All this just deflects attention in the manner of the barking dog on Ohua beach. Babylon will be the centre of the kingdoms of this world – the principalities and powers of darkness. It will, basically, be pagan in character with undertones of demonic proclivities. Citizens of the kingdom will indeed be and already are exiles from the holy city and aliens and strangers (not in the sense used in Part One) serving in the temporal enclave. This is the Babylon Peter wrote of in his first epistle and in Revelation 18: 4 saints were adjured to come out of so that they did not partake in its sins. Of course, the world had the characteristics of Babylon in his day as well.

Pope Leo XIII has been identified as an outstanding figure. According to *Don Bosco's Madonna*, October 1984 he had a shattering trance-like visionary experience in which Satan as the first revolutionary against God's programme, continued to inspire persecutions, schisms, and the ascendancy of financial and commercial powers backed by secret societies. This lined up with scripture. It was publicised in the encyclical *Humanum Genus*, 1884. The author identifies this set up as Babylon.

The journal went on to label the prevailing millennium (AD 1,000–AD 2,000), 'the second millennium of Immaculate Mary,' which is not in scripture. It also sought to show the years 1884–1984 as the reign of Satan, which was unconvincingly conjectural, except that scripture does state that Satan is the prince of this world. The author identifies from scripture that overt Satan's reign is future. Scripture indicates this will be short and calls it, "The Great Tribulation".

It is difficult for a true citizen of the kingdom even at this present time (2008) to follow a secular occupation without becoming tainted with what goes on. Extortion, deceit, and economy with the truth. But this is only the beginning. As President Reagan once quipped, "you ain't seen nothing yet!" People felt much more secure in the twentieth century, excepting during the two wars. Today they are faced with terrorism of all kinds, perverse crimes as a result of avarice and sexual promiscuity, drug addiction, and abuse of children and the elderly that was previously unthinkable. Genocides, political corruption, and pollution go largely unchecked. There is undue tolerance of religious fraud and exploitation, and a beginning of suppression of freedom of speech. All these things grieve a holy God. God is long suffering and endures the contradiction of sinners against himself.

What will happen is that evil and wicked powers will go too far. When citizens refuse to accept the mark of the Beast and decline to worship him, they will be killed along with his two witnesses. When God's patience is tried and love for his own exceeds his compassion for all people, he will be obliged to act. First of all he will bring the slain witnesses to life again. This will terrify the nations who will then glorify Him. Babylon will be destroyed in one hour by a series of divinely orchestrated disasters. Plagues, sores, polluted water, scorching temperatures, a dimmed moon, huge hailstones and blood, locusts, agonies, avenging angels, deceiving spirits, and major earthquakes which will also damage all the buildings in every nation. Most remaining people will have accepted positions in trading monopolies, and this is why the nations will mourn over the fall of Babylon. Where will the stock markets, pension funds, multi-national companies, quangos, and everything to do with infrastructures stand then?

Peter reminded citizens that they had been born again through the living and enduring word of God. They were to rid themselves of malice, deceit, hypocrisy, envy, and slander, and live good lives among pagans. The latter would accuse them falsely of wrongdoing. They must keep their tongues from evil and their lips from deceit. They must seek peace and pursue it. One need only halt and reflect that all these things citizens were adjured to rid themselves of would be prevalent in the society of that day and then acknowledge these are still rampant. Peter encouraged citizens not to fear even though they would suffer for doing good because it was better to do God's will than to do evil things. They had escaped from debauchery, carousing and idolatry. In addition to Peter's message going to the churches it was addressed to, it became part of holy writ, and it is also prophetic: it looks into the very last days.

Because the fall of Babylon seems remote it concentrates the mind to look at present realities and past history. Very recently there were two items of news indicating increased dangers. The World Health Organisation has issued a dire warning that growing air travel is causing ever increasing escalation of diseases such as AIDS, Sars, and Ebola Fever. The National Health Service Health line has a recorded advice about a plague of insect bites and stings as a result of unusually wet and humid weather. As to history, Lord Byron's poem at the commencement of this chapter is a graphic record of the Assyrian armies destruction as they encamped about Jerusalem. Sceptics will say that this was an outbreak of Cholera. That it was the angel of Death who annihilated a whole army is obvious. There is a lesson in this about the future Babylon. King Sennacherib had defied God openly. He not only said that no god could deliver Judah from his hand; he wrote letters insulting God. Sennacherib was cut down by his own sons. As to the destruction of Babylon one can make an inadequate comparison with the destruction of all the infrastructures in Baghdad in 2002 by pinpointed missiles fired from a submarine a long way away. Strategic control centres, electricity installations, oil lines, telecommunications, and water supply stations were obliterated. There is a hint in the scriptures that armies will start to turn against one another. This is what has happened with Sunnis and Shias in Iraq and it seems senseless. In the

Beast's domain it will be far worse as worship will be of demons and the bible says there will be no repentance of murders, magic arts, sexual immorality and thefts.

Reference has been made in the eighteenth chapter to the propensity of humankind to abuse this planet and it is not unlikely that this factor will accelerate the end of the present earth. It is what the wise of this world call 'the law of unintended consequences'. The international CERN laboratory in Switzerland, is conducting an experiment with its Large Hadron Collider. The object is to smash sub-atomic particles to simulate moments when the finger of God started the universe. According to a 'string theory' there are at least six other dimensions. Puny humanity is spending millions of dollars in the hope of learning more about the infinite power of an almighty God. According to the German physicist Otto Rossler, the experiment might go wrong and swallow up the entire planet! (Eaman, *East Anglian Daily Times*, 6 September, 2008).

At the sound of the seventh trumpet the mystery of God will unfold. It appears this mystery is a revelation of the things God has prepared for his people (1 Corinthians 2:9). Revelation chapter 11 takes this up. "The kingdoms of this world are become the kingdoms of Our Lord and His Christ. And he shall reign for ever and ever! O Lord God Almighty which art, and wast, and is to come; because Thou hast taken to Thee Thy great power and reigned!" (AV). Heaven will open and the Ark of the Covenant will be revealed. Some dispensationists take this as signalling the salvation of Israel. It is probably a sign that God's justice and mercy will meet over the mercy seat in the company of cherubim. A world-wide earthquake will split the great city into three parts and make a way for the armies from the east. These will go to Armageddon. That the present Republic of China is uncomfortable about teaching of the second coming is an indication of the reaction of future armies of the east when the return of the Lord becomes an actual reality. As these advance their own cities will tumble about the occupants and people will gnaw their tongues in pain. The throne of the Beast will be plunged into darkness and since this is the hub of power there will be total chaos because of complete dependence on technology. The city will become separated from its dominions. The culmination of divine judgment will be the consignment of the Beast and the False Prophet to the Lake of Fire. The great lesson of the biblical Babylon is that it teaches us the importance of righteousness. Babylon will not only be the epitome of unrighteousness, but its occupants will advocate such behaviour. They will not only be excluded from the kingdom: they would be positively unhappy within it. That is why many will still curse God even after realising who he is. Meanwhile citizens of the kingdom will hunger and thirst after righteousness, a theme to be continued in the following chapter.

21 | Christ's Kingdom

"Kingdom of Christ, for Thy coming we pray,
Hasten, O Father, the dawn of the day
When this new song Thy creation shall sing,
Satan is vanquished and Jesus is King."
C. Sylvester Horne

The major burden of Part One was to unveil a present invisible kingdom on earth that was engaged in a battle. In Part Two, conflict between the Kingdom of God and the powers of darkness of this world is still ongoing. The only difference is that this hostility is going to be brought to a head and then to a final dramatic conclusion. With no apology for repetition, The Beast will be given absolute power over the nations for a short period of time and this will encompass his downfall. He will put God's people to death for withstanding his mark. These are the privileged martyrs who will go from the earth and pass immediately into God's presence. Their song, as recorded in Revelation 15, is a valuable and important clue as to the likely outcome of the final battle. The divine intelligence conveyed to John has provided a wondrous linking of scriptures from a number of quarters. It has taken an outstanding scholar, Professor Bauckham, to unravel this and so confirm that Glory goes to God and that beside him there is none other. Just as the Israelites rejoiced in their deliverance from Egypt and in God's mighty acts, and the Passover lamb, so too the martyrs rejoice that their suffering together with The Lamb of God had brought not only their deliverance but victory over the Beast and also the conversion of the nations. The nations round about the Red Sea were both terrified and converted by the supernatural acts they witnessed; 600,000 families went out into an arid desert. With their backs to the Red Sea it suddenly parted, allowing them to cross over. The pursuing Egyptians who had gone back on their promise to let the people go were engulfed in the closing waters. Then for forty years the people were sustained miraculously in the desert by manna from heaven and water from the rock. They were guided by a cloud in the day and a pillar of fire by night. During the whole of the period their clothes and shoes did not wear out. The nations round about realised that the true and mighty God was their protector. But some of the people did not go into the promised land because of "unbelief". As the parable of the sower indicates, saving faith is not just an initial flash but a development lasting all the journey through.

When we contemplate that the risen Lord himself (by communication with John) linked The Song of Moses to events at the end of the age it provides a good lead to the outcome of the battle against Satan and his forces. The saints in heaven are singing the song. This also brings a much brighter hope for a resounding victory by our God over all the enemy. The Israelites were brought into the place of worship: the sanctuary of Jehovah. In like manner the martyrs rejoice after they too are brought into a place of worship, the very sanctuary of heaven. They were first fruits. The song in Revelation is in wording different to that in Exodus 15 but there are links through lines in Isaiah 12 and Psalm 105 that both deal with the exodus. There are also links in Jeremiah 10: 6 ,7 and Zechariah 2.10. All these scriptures make reference to the nations or the inhabitants of the earth. A summary is:

1. There is none like God. He is highly exalted
2. His wonderful works are made known. He has done glorious things
3. Make it known among the nations. Who would not fear you, O king of the nations?
4. All nations shall come and glorify his name.

Further, Psalms 96 to 100 are Messianic. In particular 96 and 98 repeat words of Exodus and Revelation:

Psalm 96
Say among the nations, "The Lord reigns."
The world is firmly established, it cannot be moved;
He will judge the peoples with equity.

Let the heavens rejoice, let the earth be glad;
Let the sea resound, and all that is in it;
Let the fields be jubilant and everything in them.

> Then all the trees of the forest will sing for joy,
> They will sing before the Lord, for he comes,
> He comes to judge the earth.

Psalm 98
> O sing to the Lord a new song,
> For he has done wondrous things.
> His right hand and his holy arm
> Have gotten him the victory;
> He has revealed his righteous acts in the sight of the nations.

Attention will be drawn later to many chapters in Isaiah that foretell of a universal coming kingdom but firstly it will be helpful to look at scriptures relating to the second coming of Christ. These are numerous and not controversial. Christians are in agreement that Jesus will one day return and judge the living and the dead. In his Olivet Discourse Jesus said he would be seen coming on the clouds of the sky, with power and great glory. He would send his angels with a loud trumpet call, and gather his elect from the four corners of the earth. After Jesus ascended into heaven two men in white stood beside his followers and asked them why they were standing looking into heaven. The men said that Jesus would return in the same manner in which he had gone. As testimony before the High Priest after his arrest Jesus said, "In the future you will see the Son of Man sitting at the right hand of the Mighty One and coming on the clouds of heaven." It is of special interest that in the Thessalonian letters Paul writes that God will bring with Jesus those who have fallen asleep. Apart from confirming that the saints return there is the news that the Father will be involved as well! Those who are still alive will be caught up. They will be for ever with the Lord. So this implies they will also be on earth with Jesus.

Before discussing the 1,000 year reign certain objections need to be dismissed. The first is that Jesus never mentioned a millennium. It is true that he did not make specific reference to it, but from heaven he had a 1,000 year reign revealed to his beloved John. He also said on one occasion that people would see "The angels of God ascending and descending on The Son of Man" (Shades of Bethel!). The most convincing thing is that Jesus set great store on the prophets. That should be good enough for us. Added to which many prophecies have already been fulfilled, so why distrust those that relate to the future? For example, Isaiah chapter 53 described Jesus as the suffering servant with great accuracy and inspired eloquence. Isaiah chapter 7 predicts the Virgin Birth. Therefore, what is the problem with Chapter 9 that says that Jesus will sit on David's throne and that the government will be on his shoulder? Of the increase of his government and peace there will be no end? The only problem is with us for not believing it. The apostles believed it and so did the early persecuted church. It gave believers a hope to cling on to. Once the church became too involved with the world and was no longer persecuted leaders turned this biblical truth into a 'heresy' and called it Chilianism. Concepts of 'The Day of The Lord' and the end times became very varied, and often subject to human wisdom and even guess work.

Today the A-millennial camp says that Jesus sits on David's throne in heaven. The Father is king of heaven according to Jesus. It is a throne of everlasting righteousness. Jesus is currently Mediator, High Priest, and Coming King. Jesus said that the Father had conferred a kingdom upon him and he waits to take this up although in a sense he is present by his spirit in believers on earth whom the Father has given to him and also those who have come as a result of their witness. So he prayed just before going back to the Father. It is not a fallen earthly kingdom because the second coming marks the end of the post-Eden creation. This must get rid of a lot of objections? The visible kingdom of Christ will be of a restored earth and an obvious progression from the invisible kingdom that already consists of new creatures in Christ and having his spirit. Jesus admitted that he was king of the Jews but that the kingdom was not of this world (i.e. this fallen world). It will be enhanced by a redeemed creation with open access to heaven of which it is an integral part. Paul wrote that the creation waits in eager expectation for the sons of God to be revealed. For the creation was subjected to frustration, not by its own choice, but by the will of him who subjected it, in hope that the creation itself will be liberated from its bondage to decay and brought into the glorious freedom of the children of God. We know that the whole creation has been groaning as in the pains of child-birth right up to the present time. So this liberation must mean that the habits of predators will cease. Those creatures that are harmful will either undergo change or else die off. One can think of insects, reptiles and rodents as becoming redundant. The angels of God will bear the traffic and this should work well seeing that there are myriads of them. It is difficult to judge which of the locations of earth and heaven can be seen as a satellite. Because the old world has been restored all the treasures and resources of the new Jerusalem will be available below: fruits for the healing of the nations and the flowing river of life to ensure immortality.

It is good to mention that there is much more in The Book of Isaiah that lines up with the Book of Revelation than set out in chapter 19. The wolf will lie down with the lamb, leopard with the goat, and the lion with the calf. Here is the restored earth. The nations will rally to

Jesus who will stand as a banner. Many of Isaiah's descriptions coincide with the revelations of Christ to John on Patmos. For instance, the day of the Lord is coming. The stars and the constellations will not show their light. The rising sun will be darkened and the moon will not give its light. One is amazed and challenged by observing the consistency with the outpouring of the bowls! Babylon will fall. The earth will be devastated and will reel like a drunkard. Isaiah foresaw the admission of gentiles even though this was otherwise a mystery not revealed until post resurrection days. He did not grasp the full implications. There are also prophecies about judgment and God's day of vengeance. Isaiah also wrote much more about the universal reign and this just cannot be swept under the carpet.

The Apostle Peter also ranked prophecy very highly. He was taught by the Lord for a period of three years. When writing of his mountain-top experience he rated even this as inferior to "the more sure word of prophecy" which we need to give heed to like to a lamp shining in a dark place, until the day dawns and the morning star rises in hearts. This star was the Messiah – the one that came out of Jacob according to the Book of Numbers, and this is repeated in Revelation when Jesus says, "I am the Root and offspring of David and the bright and morning star". He also said, "to him who overcomes and does my will to the end. I will give authority over the nations. I will also give him the morning star." In the fifth chapter Jesus is described as the Lion of the tribe Judah, the Root of David who has triumphed. The morning star is an appropriate symbol with which to herald in the millennium which is a precursor of the new Jerusalem. The morning star is seen with the rising of the dawn and can be likened to the light succeeding the darkness of the past age. Isaiah prophesied that nations would come to its light, and kings to the brightness of its dawn. The early church saw this period as 'the eighth day' or the day beyond the seven days of creation.

The Olivet Discourse was preceded by the triumphal entry of Jesus into Jerusalem as prophesied by Zechariah. Here was their king coming on a colt, the foal of an ass. This showed that a peace was foreshadowed and this is what the crowds sang about. But what was about to happen was not what the crowds expected. They were thinking in earthly terms and so was the Sanhedrin. Their creation was different from angelic beings who were transcendental. This much they accepted. What they did not know was that Jesus would pass through death to complete his Father's purpose of accomplishing the new being. Born of a virgin; taking on the likeness of flesh, Jesus was the first born of this new creation. As he said unless a grain of wheat falls to the ground and dies, it remains only a single seed. But if it dies it produces many seeds. It was explained in chapter 5 how a person can become a new creation spiritually. But we are, at the moment, mere mortals, and our bodies are corruptible. Paul explained that at resurrection we put on incorruption. Thus death is swallowed up of life. The mortal becomes immortal. When Adam and Eve were deceived by Satan they and all their descendants became subject to death. God could override this and did so in the cases of Enoch and Elijah.

In the case of Jesus there was a raising from the dead as a completion of a totally new being. Jesus was anxious to emphasise he was not a ghost or a spirit or an angelic being. He not only appeared in a bodily form but also ate and drank. As opposed to being disembodied he showed Thomas his hands and side and invited him to feel them. He could appear suddenly and disappear as quickly. When we go into death, as has been conveyed already, we have a temporary celestial body and an abode in heaven – just another part of the kingdom. It is contended that when Jesus returns to earth he will take with him all the redeemed of all ages with him as well as tens of thousands of angels. The redeemed mortal bodies will have been changed into a likeness of Christ's body. That will be for the purpose of engaging in the post-historic restored earthly kingdom. History will have been ended. This justifies deferring judgment until the consummation of the kingdom.

Two scenarios have been suggested:

1. All the redeemed bodies are raised at the first resurrection (the rapture).
2. Only those deemed to have specially attained will be raised.

In the case of 1) presumably weak citizens will be given humbler roles. In the case of 2) the weak citizens will not be raised until the end of the 1,000 years. This is unlikely on two counts. Firstly, they will already be in heaven with celestial bodies. Secondly, there would be no point in checking whether their names were in the Lamb's Book of Life at the final judgment. These would already be in it unless it is argued these could be taken out of it. Then we must face what the position will be in the eschatological millennial kingdom of those who have neither died nor been resurrected. Those of the nations who are genuinely converted are not a problem because they are part of the new creation in spirit and as will be seen will eventually be subsumed into the new heavens and the new earth, as opposed to the restored earth they are dwelling upon. The unregenerate pretending people will ultimately be destroyed along with Satan and join him. He will be cast into the lake of fire. He will not succeed in deceiving those with the divine nature as this will prove superior to the Adamic nature that failed at the beginning. This amounts to a final victory for God and an annulment of Satan's original success in deceiving Adam and Eve. It accounts for the interim incarceration of Satan in the abyss. A main attribute of the new kingdom will be peace as well as righteousness. This will be achieved because Satan is bound and the Beast and the False Prophet are already in the Lake of Fire. The citizens, with the exception of the counterfeits, will no longer have sinful natures. When they see Christ they will be like him. One can deduce what the kingdom will be like. No sickness, no crime, no death, no sorrow, no tears. The counterfeit people will be ruled with a rod of iron. Presumably movement will be supernatural and so dispense with the use of carbon fuels and mechanical means of transport. There will be no prisons, no hospitals, no disabled, no geriatrics. The economy will be divine so probably there will be no currency or financial institutions. Clothes will not wear out or get dirty. The extent to which former customs will survive is debatable but some will survive. Will these be sports, music, writing, celebrations and such like? We know that in the final kingdom the kings of the earth will bring their splendour into it.

Some have criticised the concept of an earthly kingdom and some objections are answered elsewhere. However, remarks about need for an impossible building programme to house all the redeemed of all ages on this planet ignore the fact that mansions were prepared readily enough for them in heaven. One's faith must stretch to accepting the supernatural capacity of God. There is no other way of coping with anything that goes beyond things that are seen at the present time. Our Almighty God can, and will restore and then renew the earth and fuse it with the new Jerusalem. God can take of the things that are not to nullify the things that are. This is the secret wisdom of God. No mind can conceive what things he has and will prepare for those who love him. He does reveal these things by the Spirit. God's kingdom will undoubtedly come at an unknown time in the future and this will be a corporate event. There can be no altering of what is prophesied. This leaves open the destiny of individuals so it is incumbent upon the servants of the kingdom in all generations to proclaim the good news of present enlightenment and future hope. That is to preach essentially a need for righteousness.

> When some have joy, some have sorrow,
> Some have hope, and some despair,
> How can want approach tomorrow?
> Shouldn't wealth assist it there?
> There's no room for love and pity,
> When there's greed and selfishness?

When these breed within a city
Doesn't this mock righteousness?

When the world is full of rancour
How can truth uplift its head?
What is truth but propaganda
When not seen but only read?
Shouldn't man admit his blindness
Before the Book of Life is read?
Overcome his graft with kindness:
 Resurrect the living dead!

Give them peace and give them safety,
A mortal span and life supernal.
Give them mansions, airy debt-free.
Banish hovels from time's journal.
For action do not wait for tears,
Or 'till millennial dawn,
But backward look two thousand years,
For that's when selflessness was born!
Vista by V. John Delany
(Poetic Messengers, 1998)

Isaiah speaks of a Kingdom of righteousness where, "A king will reign in righteousness." He, as is often the case, sets out all the attributes of ungodliness as a kind of reverse or negative example. "No, longer will the fool be called noble nor the scoundrel be highly respected." But that is no longer. Forthwith God's people will live in peaceful dwelling places, in secure homes, in undisturbed places of rest. He also speaks of the joy of the redeemed, and the restoration of creation. The wilderness will rejoice and blossom. A highway will be there called the Way of Holiness. People will enter Zion with singing. Gladness and joy will overtake them and sorrow and sighing will flee away.

When that which is perfect comes tongues (plural) will cease altogether. The world's punishment for building the Tower of Babel will no longer be in force and the attendant confusion will be a thing of the past. God will be worshipped with one unified voice and one common language. Knowledge will pass away – that is to say the quest for further knowledge. The Fount of all Knowledge will provide completeness.

22 | The New Jerusalem

"Now the dwelling of God is with men, and he will live with them. They will be his people, and God himself will be with them and be their God." *Revelation 21: 3*

"Progress should mean that we are always walking towards the New Jerusalem. It does mean that the New Jerusalem is always walking away from us. We are not altering the real to suit the ideal. We are altering the ideal: it is easier." *G. K. Chesterton, Orthodoxy*

God will bring his purposes about at the pleasure of his own will. He enjoyed walking in the garden with Adam and Eve. Following their disobedience they hid from God as he walked in the garden in the cool of the day. Then their communion with God was sundered. After this there were sporadic encounters with "The I Am", such as Enoch, who walked with God, and Abraham as already described. Then there was the burning bush encounter with Moses. A more corporate and terrifying event was when God appeared on Mount Sinai. Moses ascended the mountain and the people remained assembled at the foot. The Lord came in a dense cloud. There was thunder and lightning, and a very loud trumpet blast. Everyone in the camp trembled. The mountain trembled violently and was covered with smoke as the Lord descended on it in fire. Here the law was given to the children of Israel who could not keep it any more than Adam could. The sight was so terrifying that Moses said, "I am trembling with fear." But departing Christians do not come to this awful mountain. They come to Mount Zion, the heavenly Jerusalem, the city of the living God. To thousands upon thousands of angels in joyful assembly, to the church of the first-born whose names are written in heaven. After this God sent his own Son and he was in Christ reconciling the world to himself. Jesus was the express image of the Father. We beheld his glory as of the only begotten of the Father full of grace and truth. The Son completed his Father's plan through death and resurrection as the first-born from the dead of a completed new creation. This is reiterated in order to draw a sharp contrast between the heavenly celestial bodies possessed by departed saints and Christ's resurrected body which will be given to all raptured citizens and all heavenly citizens on the day of the Lord. Heavenly citizens will consist of all departed ransomed people who entered the spiritual kingdom during their mortal lives as well as righteous Jews of past ages who are part of the Israel of God. There was later no distinction between Jews and Gentiles. Then one will either be in or out of Christ.

In the previous chapter the sojourn of all the faithful on a restored earth was embraced. We have now to consider the final development of when the New Jerusalem comes down and either docks to a new earth or else remains just immediately proud of it. The Expression 'above' is avoided because the world is a rotating and moving orb. The statement, "behold I make all things new," shows that there is a further transformation of the earth. The most obvious conclusion is that the ultimate renewed earth will take on the atmosphere of heaven. Old things will have passed away as symbolised in the figure of the elimination of the sea. The sea must mean the plasm, from which God fashioned the first world and would become obsolete. In that God will dwell with men, there is significance in the word 'dwell'.

There were other ways in which God came to men in the past. It was by his glory filling the tent of meeting or later the temple. On the day the tent of testimony was set up in the wilderness a cloud covered it and it looked like fire. In king Solomon's day the Shekinah so filled the tent that the priests could not perform their service. When the ark was brought to the newly constructed temple the cloud of glory once again was so intense that the priests could not perform their duties. Eventually the Shekinah left the temple and according to Jewish writings remained on top of a mountain for many days and was heard by many to be lamenting. This glory, this Shekinah of God, is what lights the Holy City. Isaiah prophesied, "Arise, shine, for your light has come, and the glory of the Lord rises upon you. Nations will come to your light, and kings to the brightness of your dawn." Paul encountered the divine glory on the Damascus road. He saw a light from heaven, brighter than the sun, blazing round him and his companions. They fell to the ground. On this occasion the ascended Lord spoke audibly to him. So bright was the glory that Paul suffered a temporary loss of sight.

John's remarkable and vivid description of the architecture and landscaping of the new Jerusalem are the best possible within the limitations of our nomenclature. Precious stones of the highest quality, translucent gold roads, rivers and canals. This is highly romantic but totally inadequate as far as the reality is concerned. Its four walls, with three gates on each side from which the saints go out (Ezekiel 48:30) and the nations enter (Rev. 21:4), are a confirmation of what is being presented apocalyptically about what is beyond comprehension. Paul said that the Jerusalem that is above is free, and she is our mother. He also said that our citizenship is in heaven. And we eagerly await a Saviour from there, the Lord Jesus Christ, who by the power that enables him to bring everything under his control, will transform our lowly bodies so they will be like his glorious body. This, surely, confirms the view that the redeemed celestial bodies are a temporary habitation of the spirit until the rapture upon which raised mortal bodies will be transformed into a resurrected body for the purpose of dwelling firstly on a restored earth, and then eternally on a conjoined renewed earth and a heavenly Jerusalem.

Most children's stories used to end by saying that they lived happily ever after but this was a kind of code word with which to say "finish". There is a sense in which the new Jerusalem is where people live happily ever after but it is far from "finish" or a place of eternal rest. Because it is not possible to describe heaven in finite terms, writers have struggled to tell us exactly how its inhabitants will be occupied. Answers have been attempted in this work. John Bunyan presented his view of the heavenly city as a dream and this was both honest and beautifully done. Somehow imaginary descriptions provide blessing through the spirit. Isaiah did well to compare peoples' present state of Deserted and the land as Desolate with heavenly counterparts of Hephzibah (my delight) and Beulah (married). Just as the early persecuted church found enormous consolation in thoughts of eternity so too did the Negro slaves transported to the New World. They had virtually no hope of deliverance from gross mistreatment in this life. However, their songs are amongst the most deeply moving and inspiring. "Swing low, sweet chariot, coming for to carry me home" . . . "If you get there before I do, tell all my friends I'm coming too." "Deep river my home is over Jordan."

The older one gets the greater is the tendency to start thinking about the prospects of dying. This ought not to be the case because we do not all die after the same number of years or because of degeneration. There are such things as accidents and violence and cruel viruses. A correct attitude is to be intent on serving the Lord while we are alive. The less that this world means to us, as in the words of a hymn writer that we are "weaned of this passing show," the more we will look forward to going over Jordan. There are some cities or towns that are either set on a hill or are adjacent to much lower ground. In both cases it is possible to get a view of the whole conurbation from quite a distance. It is too far to observe movement or to hear noises. There is something that touches the senses that is moving and mysterious. Of course, once in the place it is very much more ordinary than imagined, and, therefore, rather disappointing because it is earthly. The new Jerusalem will be more impressive than any first pleasant encounter because it will be against the sky and descending towards observers rather than they going towards it. And skies are magnificent especially in flat places and over the sea. But the greatest difference will be upon entry. Apart from the unimaginable beauty and tranquillity of the scene, God himself will be dwelling within this mobile heavenly spectacle. Ideas of crowns, white robes, and continuous praising, are substitutes for superabundant and supernatural preparations.

This will mean that in coming down from heaven God will get involved with governance of his particular people and the nations at large. Heaven and earth will pass away, that is the present heaven and earth, but God's word will never pass away. In fact God's creative and living word was the voice that accompanied the trumpets and which is also probably the river of life. It was a voice that announced, "now is God dwelling with men!" The new Jerusalem and a new earth combined is a continuum of the first 1,000 years of blessedness. As it comes down from heaven it is as a bride adorned for her husband. The celebration of this tremendous event is called the marriage feast of the Lamb. When Jesus spoke of drinking of the fruit of the vine with his followers anew in the kingdom it would seem that the consummation of the kingdom would amount to the greatest rejoicing. Of course, all this language is clearly metaphorical. This is why it is so difficult to convey fully what future joys are going to be like.

On account of such limitation it will be appropriate to reflect on why this reward for the Lord was so meritorious. The father had placed a great burden of responsibility on him. He was to eradicate the blight of original sin and defeat Satan. He was well armed for this task by the Father. Adam was of the earth, earthy, but Jesus was the Lord from heaven. The theological key to this is his virgin birth. He was not born of the will of man but of God. This means that he took upon himself our humanity and emptied himself of an equality with God. But his superiority over Adam's race was that he did not have a fallen Adamic nature. This will prove to be God's extraordinary weapon at the end of time. By his substitutionary death Jesus atoned for sin (singular) and by his death and resurrection made it possible for Adamic beings to receive a divine nature like his own by the power of the Holy Spirit. The forgiveness of sin is a judicial finding. It is a reprieve from a second death. But a sinful Adamic nature will remain with persons so long as they inhabit a mortal body.

So these two natures conflict and are contrary one to another. The sinful nature must be reckoned dead daily. One is rid of this through mortal death and comes into a glorious freedom. It is the great triumph of God and his Christ that this new creation assists in the final defeat of Satan and the flesh, bringing about the first resurrection, a millennial reign, and the new Jerusalem. An ability to be completely Christ-like is the ultimate gift of freedom by God to his people. That they have come through much is recognised and rewarded. The rider on a white horse carried a title – King of Kings and Lord of Lords. His garments were splattered with blood and this was the blood of the martyrs not his own. In a very subsidiary way they share in his victory over Satan.

Some very important matters will be resolved at the close of this time. Those people who were unregenerate but quiescent during Satan's confinement in the abyss will prove to be as the sand of the sea, that is countless. They will be destroyed along with the released Satan who has deceived them and all will go into the Lake of Fire. This will happen almost co-terminously with the raising of the rest of the dead who will stand before the Great White Throne of judgment. There will be a finality in the sense that all evil is permanently done away with. All things will be made new. Those who hungered and thirsted after righteousness and mourned about the failure of their sinful natures in the old world will know that the blessings they enjoyed in the millennial kingdom have been sealed in perpetuity. God's word has told them that as the new heavens and the new earth endure before him so will the citizens' name endure from one new moon to another. All mankind will come and bow down before Him. These will be the converted nations who will enjoy all the supernatural healing provisions of the leaves of the special tree in the holy city.

In this world one can access the truth of God and also encounter what is counterfeit. The truth of three persons in one God is offset by a demonic trinity of Satan, the Beast and the False Prophet. The stones connected with the occult zodiac are in the reverse order of the stones in the holy city. And so these will appear in the city of Babylon and its dominions. It can be assumed that this is a counterfeit of the new Jerusalem and the renewed earth. When the bible states that Christ will be given dominion it refers to his future reign (he is presently our intercessor) over all the nations and later from the heavenly city, now alongside the renewed earth. The government will be upon his shoulder and he will be called Wonderful, Counsellor, The Almighty God, The everlasting Father, The Prince of Peace. It was as the Prince of Peace that he entered the earthly Jerusalem "lowly" sitting on a colt the foal of an ass. That underlined his character as the servant king and added to the lessons in humility and service he tried to instil into his followers. They were always slow of heart to respond or to believe all the prophets had foretold. However, in the eternal round Jesus, as an unchanging model of compassion and servitude, will be fully supported by his new creatures. It may prove hard to enthuse about this in one's present spiritual state: a bit like the half-baked Ephraim in Hosea's prophecy. We still groan for the full redemption of the first resurrection. Then when we see him as he is it will be a delight to dedicate ourselves to the service of righteous government and the care of the nations. Of the increase of his government there shall be no end, upon the throne of David, and upon his kingdom, to order it, and to establish it with judgment and with justice from henceforth even for ever and ever. The zeal of the Lord will perform this. All this is but a flavour of what citizens will encounter in the new Jerusalem – a new heaven and a new earth wherein dwelleth righteousness.

PART THREE

Summary and Conclusions

23 | A Review: Mystery, Majesty, Immanence

"O loving wisdom of our God!
When all was sin and shame,
A second Adam to the fight,
And to the rescue came.
O wisest love! That flesh and blood,
Which did in Adam fail,
Should strive afresh against the foe,
Should strive and should prevail."
   Dream of Gerontius, J. H. Newman

Part Two of this book is dealt with in the Summary and Conclusions first not because it is of the most importance. It is so ordered because it will be recently in reader's minds and also because this will leave the end space for review of Part One, which is of more pressing and primary importance in contemporary lives.

Part Two: The Coming Kingdom

"Man proposes but God disposes."
   Imitation Christi, ch. 1 xix, Thoma à Kempis

The author cannot claim to be completely right when deductive reasoning has been applied to facts, but trusts that, overall, readers will find the study to be constructive. It is thought there will be a period when Jesus will reign visibly over a kingdom on a restored earth after the whole fallen creation is redeemed. If a different view is taken that the second coming will herald a final end, including harvesting of the earth, and immediate judgment followed by Jesus and the redeemed being permanently in heaven, then it would mean that the nations will remain substantially unconverted and that the powers of darkness are overcome by one forceful act of vengeance. It would be difficult, under such circumstances, to see what sort of a kingdom Jesus would at some stage hand over to the Father. Jesus made it very clear that the prophets were to be believed. When coupled with the Psalms there is a strong case for a redeemed intermediate dominion of one kingdom on earth. This will be conferred upon Jesus by the Father and will ultimately be merged with the descending new Jerusalem. Once having cleared the decks it was possible to shed new light upon end time possibilities and a coming visible kingdom.

The scriptures leave us with some uncertainties and a need for conjecture. Reference to the Old Testament gave added weight to some possibilities. A remaining tension is what degree of success will be achieved in seeking to win the nations. Why did Jesus say virtually nothing about this when Revelation, the Psalms, and the Prophets gave a number of indications? The superb, divine mind of Jesus confined this tension to what was really important, namely, to warn the unrepentant about what could be in store for them, and to enthuse his followers to pursue a more urgent course of witness even to the point of martyrdom. It is hoped that readers found the expansion of eschatology informative and interesting. Everybody tends to be curious about the future but predictions are not as vital as, say, the gospel message, but it is educative when inquisitiveness is satiated and debate is stimulated. It must be allowed that the Olivet Discourse was given immediately prior to the Lord's arrest, trial and passion, about which he had foreknowledge. There was no time for niceties! In any case he was also aware that he could communicate by the Spirit with his beloved John after his ascension. You will have gleaned that the author is optimistic about God's glorious final victory, which will, judging by the revealed character of God, benefit the inhabitants of the earth so that they will glorify him and sing, "Our God reigns."

The suggestion that the millennium will run on a restored as opposed to a renewed earth is novel. However, it is supportable both from scripture and from anticipating a total reversal by the Almighty Father of any triumphs Satan may have achieved in the interim over the weak Adamic flesh of the fallen creation. An important realisation is that the Devil's second opportunity to subvert the nations on release from the abyss will be God's clever design. Together with Gog and Magog; Satan will fail to deceive or penetrate the defence of the divine nature in the way he did with the Adamic race. Positing that there will be a restored earth overcomes claims that a present earth is not fit for occupation by incorruptible bodies, and fits neatly into the author's chronology of a later final renewed heavens and earth when the new Jerusalem descends to the earth and the glory of God covers it as the waters cover the sea, and righteousness dwells in it.

So to recapitulate the chronology, the Beast will so persecute witnesses that God will wreak vengeance on him and his fellow miscreants. Genuine believers will either be killed by the Beast for refusing his mark or else be specially protected as was the case with the first Passover. They will however be taken from the earth to escape God's vengeance and to rejoice in heaven at the triumph over the powers of evil, their own deliverance from it and the conversion of the nations. All members of God's kingdom whether in heaven or on earth will take part in a first resurrection. It would be illogical to suggest otherwise because all departed citizens will already be in heaven with celestial bodies. It is very difficult to place the harvest of the earth and the final resurrection and judgment in a completely convincing sequence. The inclination was to have these placed at the immediate time of the Parousia. However, it seems on balance that all these events must be put at the very end because the unregenerate dead are not raised until the end of the 1,000 years. As to the reaping of the earth, this would seem to tie in with the parable of the weeds which are left to the end lest good plants are pulled up by mistake.

It is noted that a grain harvest and a grape harvest appear to relate to two different types of person. Grain is sorted from chaff and put in the garner whereas grapes are trodden out. When this is the scenario it gives some hope that people who never heard the gospel or those who were not positively evil might be spared from the Lake of Fire. This could mean that some people will be gathered into God's garner and not be trodden out with the grapes of wrath. A single final judgment appears to be most likely because it appears that the rebellious kings, who side with Satan in his last fling, will be judged along with the archdemon himself. The new Jerusalem will descend to the earth outside of finite time, and it is comforting that the nations as such still exist and will come in and out of the eternal city. God himself will tabernacle with humanity. He will be their God and they will be his people.

According to the writer of Hebrews, "no longer will a man teach his neighbour, or a man his brother, saying "Know the Lord," because they will all know me, from the least of them to the greatest." This was referring to the High Priestly office of the Lord at the right hand of the seat of the Majesty in heaven. At the bottom line this means not being beholden to any man made system. The Kingdom is truly united, unassailable, and can be opened with the keys of the gospel, and being born again. That is partaking of the divine nature. The mystery hid from the ages and now revealed is, "Christ in us, the hope of glory." Unless one is born again of water and of the Holy Spirit it is not possible to enter the kingdom.

It is submitted that little thought has been given to the destiny of departed believers. Spare a moment first of all to contemplate the sheer and growing volume of saints entering glory from age to age. Then consider the manner in which they will come back. Thousands upon thousands will accompany their Lord at his second coming. Then with an eye of faith see them perfected by him with unassailable characters. This is because they will have seen Him face to face and been changed into His glory as by The Lord the Spirit. They will be like Him. They will be conformed to His image. The victory of God's Christ and his followers will be total and utterly complete. Maranatha! Our Lord come! These are not empty words and one hopes to be forgiven for those years in which the word of God was parroted in a glib fashion, or for even later when the fullness of meaning could have been plumbed sooner and allowed to produce a greater sense of urgency.

Can joy in the Holy Spirit be explained? Well it is like what the martyrs will experience when, clothed in white garments they will sing the Song of Moses. "King of ages. Who will not fear you, O Lord, and bring glory to your name? For you alone are holy. All nations will come and worship before you, for your righteous acts have been revealed."

> God moves in a mysterious way
>   His wonders to perform;
> He plants his footsteps in the sea,
>   And rides upon the storm.
>
> Deep in unfathomable mines
>   Of never failing skill,
> He treasures up his bright designs,
>   And works His sovereign will.
>
> Ye fearful saints, fresh courage take;
>   The clouds ye so much dread
> Are big with mercy, and shall break
>   In blessings on your head.
>   *William Cooper, 1731–1800*

Part One: God's Present Kingdom

"Faith without works is dead." *James 2:20*

When Erasmus wrote a damning condemnation of the state of the professing Christian church in a dialogue, Martin Luther expressed concern in his Weimer Ausgabe. "It is so agreeably, learnedly and wittingly put together, that is, so thoroughly Erasmian in fact, that it compels one to smile and jest on the subject of the faults and misfortunes of the church of Christ, which, however it is every Christian's duty to deplore before God in deepest grief".

When the prophet wrote, "He who sits in the heavens shall laugh," surely he did not mean that God would laugh at us? Obviously not, because it is not in God's nature to regard the faults of Adam and his progeny as amusing. One hopes and trusts it is to be a future victorious laugh in eternity when the divine purposes have been accomplished. A likely answer will be given later. When God created the universe, and the world in particular, his finished work on the sixth day was good. Prospectively Adam could look forward to a kind of super Utopia in perpetuity. Then it all went wrong. In the foreknowledge of God (which does not negate human responsibility) there was a solid Plan B. God needed to intervene in history. So the kingdom of God came into history when God sent his Son by supernatural means into our world. After being anointed with the Holy Spirit at Jordan when 30 years of age, Jesus exercised a three-year ministry of preaching, teaching, and training accompanied by acts of power.

In him the kingdom was "at hand" or imminent. By his atonement for sin on the cross of Calvary and a vindicated resurrection from the dead Jesus defeated Satan and became the first-born of a new creation. The kingdom became open to humankind for the first time. It became patently clear to believers that for ever afterwards they could be born anew to a spiritual life. However, in spite of this initial marked emphasis on a present spiritual kingdom by Jesus and those he sent out, and in the instructions given by him during the forty days between his resurrection and ascension, this then present truth receded.

Luke and Mark included some exact resurrection words of Jesus in their accounts of The great Commission to go into all the world and preach the gospel, but recorded nothing about what he said about the kingdom. Luke, the writer of Acts, tells that Jesus spoke first of all about the kingdom. Absence of detail may be due to a difficulty of writing about invisible spiritual things. The things that are unseen are eternal, whereas things that are seen are temporal. Paul's missionary travels were dominated by the preaching of the kingdom. However, the subsequent growth of the visible church, perhaps understandably, resulted in increased preoccupation with the gospel, and formalities, and tradition. As less and less importance was attached to appreciation of a present kingdom, the raison d'etre of empowerment for works of service also became obscured: especially as a means of combating powers of evil in this world of darkness. It is not always realised that the kingdom is present in whatever sphere the righteousness of God holds sway. Citizens within it, by means of grace, faith, and the word, prayer and spirit empowerment, contribute to the forward advance of God's purposes. Satan is their major adversary who tries to frustrate them by temptation and deceit. He can never prevail against the kingdom itself. He has his own disciples in the shape of false prophets and teachers who are involved in the operation of counterfeit power. Ultimately God's kingdom will prevail against Satan and his kingdoms of this present world.

A major objective of this book is to give greater exposition to a teaching about a present kingdom with a view to inspiring and motivating citizens and of restoring such teaching as an utmost priority. This is the place that Jesus gave it amongst the other wonderful works of God. Another objective is to point out the longitudinal character of real faith. Yes, people are justified by faith, but this is not something to be ticked off by mental assent: it must be ignited by the Holy Spirit, and the truly believing one is kept by the power of God.

Citizens can, and do, sow to their sinful Adamic natures. When such sowing is submitted to, instead of walking in newness of life, it hinders the progress of righteousness and gives place to sinfulness. It is only as Christ's people go through mortal death that their sinful natures leave them. They are for ever after Christ-like and impervious to both satanic deceits and inclinations of the flesh.

An answer must be given to a question of whether belief in the kingdom of God is a Delphic delusion or radiant reality? The temple of Delphi was the centre of a mystery religion. Once initiated, entrants had to swear to secrecy under pain of death: they were deluded psychologically. But the presence of the kingdom of God as a present reality is not a secret to be withheld from anyone. As has been explained it is simply hidden to many for a number of reasons. Neither is it a mystery or mythological delusion. Following the habit of the lady who always goes to the end of a book first (see chapter 6), it can be seen that Part Two brings this question into very marked contrasts. A millennium (at Christ's second coming) was accepted completely by the Apostles and the early church for a period of 300 years. That would have been the era of small persecuted minorities. One can imagine the blessed hope this doctrine brought to these beleaguered people. Then as the church began to enjoy the patronage of a quasi Christian empire, namely that of Emperor Constantine, this biblical truth became labelled as a heresy under a title of Chiliasm. And thus things remained until increasing schisms in the visible church made it convenient to resurrect parts of this subject by identifying either opponents or accepted tyrants with apocalyptic characters.

This has already been explained but to add two more fantasies: Pope Gregory VII was seen as the literal Anti-Christ and Napoleon Bonaparte as Apollyon! All these suppositions amounted to delusions and so placed the subject of eschatology in the bracket of the field of imagination. It then became a plaything for academics and poets, and an intellectual fiction for the public at large. Re-emergence of the doctrine as a biblical reality was shown, unfortunately, to be subject to a variety of interpretations. Even when teaching belongs to one camp or another the core of truth (e.g. a second coming) was clear enough. In its purist form it is the greatest challenge to mans' established authority whether this is in religion or secular government. The concept of a divine overall government of the world of which the Lord is king causes consternation whether it be a totalitarian state like the great nation of China or "the best democracy that money can buy", as a Cambridge academic has described the United States of America . The Chinese permit the existence of Christian churches provided these are registered with the state and do not teach the second coming of Christ. Both in North America and in Britain a President and a Monarch respectively confess Christ. This is comforting on the contemporary scene because God was prepared to spare a city if only a few good people could be found. They could not.

Another oppressed and beleaguered people found great consolation in thoughts of eternity. The Negro slaves in the Southern States left a legacy of this in their wonderful spirituals. (Swing low sweet chariot, coming for to carry me home . . . if you get there before I do, tell all my friends I'm coming too. Deep river, my home is over Jordan.)

It has been claimed that a universal acknowledgment of the existence of a present earthly, spiritual and dynamic kingdom fighting against the powers of darkness would revitalise and recapture this real truth. Furthermore, it would result in a greater motivation of citizens to work zealously in taking the gospel to the world. Potentially, it could also be an avenue for promoting unity amongst the visible churches. It would require a great deal of humility on all sides to achieve such an objective. This is because the visible churches as such are not synonymous with the true kingdom although a majority of its citizens are drawn from them and they are repositories of the Word of God here below.

Giving greater prominence to the kingdom will create tensions. It may be seen as subverting sectarian interests in a corporate sense. However, as a means of encouragement to citizens it is suggested that a re-examination of scripture in the light of a recaptured fuller vision will elicit much deeper meanings. For example the Lord's Prayer can be seen as more of a present reality than a future hope. When Jesus said that those who set their hand to the plough and then looked back were not fit for the kingdom it makes complete sense when this statement is seen in the context of a present reality. Otherwise it is very difficult to understand! Evil is also brought more into the open and shown as a force that is in conflict with light and righteousness (and vice versa). However, there are warnings that naïve Christians can be exploited by exaggerations about demonic forces. For example, Paul was more concerned in his later ministry with nasty human characteristics whether revealed at personal or corporate level. Hymenaeos did him great harm at one point, believing a general resurrection had already taken place. Everyone in Asia turned away from him, including Phygelus and Hermeogenes. He did not attribute these problems to demons.

In the later chapters it was sought to re-anchor futurity to the scriptures. In doing so there was a de-bunking of all the hair-brained ideas formulated over the centuries that were little better than the admitted fictional fantasies of present-day writings. These latter range from science fiction, and the Harry Potter novels to The Da Vinci Code. It is worth noting that there have been accurate historical publications of an academic nature that contributed to knowledge without dealing with spiritual truths. A good and laudable example is *The Pursuit of the Millennium: revolutionary millenarians and mystical anarchists of the middle ages*, by Professor Norman Cohn. By rejecting the fictional aspects of eschatological propositions the potential readership of this book may well have been reduced. However, it is trusted that this will not affect their quality. The stark reality is that the more confusion there is the more Satan and his followers will be pleased.

The most practical and immediate subject is the existence of a present invisible dynamic kingdom in the here and now which calls for attention and action. A church member said that having heard a sermon on eschatology he was disappointed that not very much was said about the king. It is, therefore, good to pay special tribute to the king. What a thrill it will be to meet Jesus who as well as being king, is also God and the Son of Man. Some earthlings would find it more interesting if Jesus had proved to be some super alien from another planet who had been beamed down, or perhaps some kind of a spook. God is too practical to be involved in anything that is surreal or phantom like. When Jesus rose from the dead and appeared to his followers he was not a spirit or a ghost but very much a material being who ate and drank and conferred audibly. This king, for such he is, came into the world in the most menial manner despite the fact that he had a glory with his father before the world began. Nobody ever spoke with such authority as he did about eternal things past and present. "Before Abraham was, I am."

"You will see the Son of Man sitting on the clouds of heaven and coming in glory." Vitally the risen had an incorruptible and glorious body.

This is also a pointer to the nature of all resurrected bodies of believers in the future. To gain an understanding of the relationship of the kingdom to those who enter it, to those who serve it faithfully, and those who obey the king, one only has to examine its nature. Therefore, such things as sacraments, religiosity, charismatic experiences, pilgrimages, whatever their merits or lack of them, are as nothing when compared to righteous and obedient living. The Kingdom of God is righteousness. It is not the purpose of this book to attack the sincere beliefs of others but to draw attention to the dangers of possible distraction. A large number of citizens or otherwise religious people are influenced by denominational backgrounds but each is an individual who sometimes thinks outside of what bounds he, or she, is committed to. All go into death as individuals and will answer as such.

The following adds weight to this contention. An extra-ordinary late discovery has provided a great impetus to the thrust of this book, and demonstrates the accuracy of the previous paragraph, and particularly to both the claim of a present earthly kingdom and the contention that acknowledgment of this would be a means of promoting greater unity. Twenty-four years ago the author gave a series of public lectures on Catholicism in Southend, Essex. He was pleased but rather put on his toes at the presence of a Catholic Parish Priest and some of his flock. The object was to explain what the Catholic faith was about. The approach was to set out belief based on Catholic teaching. A minor lesson was that the Catholics present were largely happy with what was said, while the Protestants were quite disturbed at some of the facts. This is indicative of two mindsets. After the final lecture the priest said he had thought the speaker might be an aggressive person like Dr. Ian Paisley (that was in 1983) or someone who had mugged up a few facts from a book. However, he was impressed not only with the accuracy of the lectures but with the kind manner in which these had been delivered. There was an amicable disagreement over one point but it transpired that the priest held a slightly different view on this than was Rome's official line, proving the value of private conscience.

This priest presented a book to the author with the thanks of the Catholic community. And this book has now hit this author forcefully. It is an edited version of the diaries of Pope John Paul II. In about 1993 (about the time of his Marlowe Lectures) the author discovered that the editor was the priest who presented the book to the author. So he was no ordinary parish priest. Now when a Pope speaks ex cathedra he is reckoned to be infallible. A son of his church he is bound by his loyalty and commitment. That is why the respected and loved late Cardinal Hume adopted a similar 'up the organisation' stance when he told defecting Anglican clergymen they 'must accept the whole menu' in order to come over to Rome.

So what is so stunningly different about the diary? Writing as a man this Pope stated, "The kingdom of God is righteousness and peace and joy in the Holy Spirit." He also said that "division dims the proclamation of the kingdom of God. Whenever we recite the 'Our Father' this can be an intention to pray for unity" (entry 23 January 1980). On 22.3.64 He wrote about tolerance and conscience. "An individual can become convinced that he is right and this can be a very profound problem of his inner life. A problem the church respects (?) (Author's question mark). From the very beginning, from the writing of the Apostles, we come across clear traces of respect for the conscience that is in error but at the same time is convinced he is in the right. And in this lies the problem of conscience". It will be seen that this Pope privately shared the author's convictions about the truths of a present spiritual kingdom, and that he thought this was an important key to Christian unity. Moreover, the 'Our Father' was a, confirmatory, unifying prayer. It also seems to the author that the diary shows a sympathy with private conscience, an attitude that Roman Catholic doctrines are at odds with.

In fact Rome considers every digression from its own tenets to be error. This comes over strongly in the Vatican II documents, which state, "The church previously treated error with the utmost severity. Now she uses the medicine of mercy by demonstrating the validity of her teaching rather than by condemnations". It is a pity that Pope John Paul never got around to putting all his private views into ex cathedra statements. He would have electrified the Christian world if he had done so. There was an interesting comment in one of his Papal encyclicals – *Veritatis Splendor*, "The reality of the Kingdom is referred to in the expression 'eternal life,' which is a participation in the very life of God. It is obtained in its perfection only after death, but in faith even now a light of truth, a source of meaning of life, an inchoate share in the following of Christ. Indeed Jesus says to his disciples after speaking to the young man: Every one who has left houses or brothers or sisters or father or mother or children or lands for my name's sake, will receive a hundredfold and inherit eternal life (Mt 19:29.)" This is very encouraging because in effect it is an endorsement of the evangelical gospel which proclaims the truth of an indwelling Christ-life that can be received as an entry into the present kingdom, and also that in this life it is in conflict the Adamic nature. But the encyclical also underlines a very sad finding in this book made more vivid by linking President

Roosevelt's difficulty over mixed information with a barking dog on a Pacific beach. The Papal nugget is hidden amongst very controversial statements that are likely to put many off the sentences highlighted by the author. For example the encyclical goes on to claim that when Mary stood at the foot of the cross with the beloved John she asked, together with Christ, forgiveness from the Father for those who do not know what they do. This cites Luke 23. 24, but this scripture does not say this. Scripture is all-sufficient and truth cannot stand upon what is not there!

Human Life International recorded in 1994 an article from the leading Roman Catholic Newspaper *The Universe*. It stated that sixty theologians in Quebec Province had written to their bishops to say they will not accept *Veritatis Splendor*! Continental European theologians are of the same mind, many of them, as the battle rages in theological European literature. "What could better demonstrate the crisis in the Church than the rejection of *Veritatis Splendor?*

Of course, if they accept that encyclical and the catechism they must take back all their dissent and all their false teaching of years. What a mess!"

Was it the nugget identified by the author that caused the stir, or was it the misuse of scripture? It appears that it was neither. Writing in *Christian Order*, a Catholic journal in 1993, Gerald Warner made it clear that there had been acclamation of "an effusion of Marian devotion." It appears that the main disquiet was about the strong reiteration of the Church's teaching on birth control, homosexuality, and sodomy. It presented many people with 'difficulties'.

Warner continued, "Of course, it will. So does the Ten Commandments." This was a case of the temporal things that are seen outweighing eternal unseen issues.

A strong conservative lobby is still dominant in Catholicism because *Veritatis Splendor* contained thirty four anodyne citations of *Gaudivin et Spes*, the council's pastoral constitution of the Church in the modern world, whereas there were only seven references to *Dignitatis Humanae*, the declaration of religious freedom. It is salutary that in Warner's opinion the latter encyclical was a controversial document, "which traditionally minded Catholics hope will eventually be abrogated." He added, "Dissent now signals infidelity, by axiom that, Rome having spoken, the issue is closed: Roma locuta, causa finita."

Therefore, the nugget in *Veritatis Splendor* was akin to that little signal missed by President Roosevelt. Continuing oversight might prove more disastrous on a spiritual plane than the cataclysmic happenings at Pearl Harbour were in history. The author's pessimism about the prospects of unity seems to be confirmed.

This is despite the early Galilean ministry of Jesus consisting of preaching the Kingdom of God and giving instructions about life in the kingdom. This is why the Beatitudes are so sublime. The kingdom was for those who were poor in spirit. That had nothing to do with worldly poverty or religious background but with humility of spirit. By way of illustration two of the beatitudes are now selected and related to Parts One and Two, respectively. Part I (fourth beatitude) blessed are they who hunger and thirst after righteousness for they will be filled. What does this mean? It does not mean those who suffer as a result of persecution. This is covered by the last beatitude when this is the fourth. The Catholic Knox and Jerusalem Bible translations state respectively hunger and thirst "for what is right" and "for holiness." A young man asked Jesus which was the most important commandment and was told that it was to love the Lord your God with all your heart, and with all your soul and all your mind and with all your strength. The second was to love your neighbour as yourself. When the young man agreed Jesus observed that he was not far from the kingdom of God. As to part II: blessed are the meek for they shall inherit the earth must, surely relate to the millennial reign because there is no prospect of the meek getting a look in when it comes to this present world's rat race. The kingdom never changes as to character and attributes quite regardless of changes in location or environment. Therefore, the fourth beatitude applies to the ethics of both present and future periods of kingdom life.

All right-thinking believers would like to see a unified, righteous, visible single church run on the principles of the teachings of Jesus. They would desire to see the conversion of the nations. Scripture states that at the end Jesus will present to himself such a church. Just how far God's inexorable purposes will be achieved before the end times is not known. However, despite understandable eagerness, it is known that final divine intervention will become necessary on a large scale. An angel will be sent out with the gospel after the bringing back to life of the two witnesses.

Therefore, a less favourable scenario should be considered. If visible churches remain as they are or continue, generally, to spiral farther away from the bible, the one true invisible church will continue and, hopefully, be spurred by the state of the world. What did Jesus mean when he said to the Samaritans, " a time was coming when you will worship the Father neither on this mountain nor in Jerusalem? . . . the true worshipers will worship the Father in spirit and in truth, for they are the kind the Father seeks. God is spirit and his worshippers must worship him in spirit and truth."

It might be asked if there is some interim alternative. The chance of a further reformation in one or more of the mainstream churches is slender. The present tensions in the Anglican church could lead to a split in which a wing might go back to biblical basics. Historically, all attempts at returning to apostolic doctrines have only met with partial or temporary success. Disappointingly, the perpetrators of distractions finished up on top. Examples are: in Catholicism the Ultramontanes and in Protestantism the liberal school. Many fine men have tried to urge churches to go back to biblical simplicity, and others have laboured sacrificially on mission fields, demonstrating by actions what true Christianity is. Not all of these had a full biblical vision. In Catholicism examples are: Thomas a Kempis, Bishop Challoner, Fathers Lupino, Serra, and Damien, Ignatious von Dollinger, the Jansenists and the Gallicans. In Protestantism examples are: Martin Luther, Frederick Schleiermacher, Wycliffe, Tyndal, John Wesley, William Booth, Hudson Taylor, Wilberforce and Bishop J.C. Ryle. The latter had a perfect understanding of the true church.

This church can be summarised as being comprised of all those who have experienced a genuine second spiritual birth confirmed by evidence of a changed life. The prophet Isaiah gave examples of confirmatory behaviour in the fifty-eighth chapter of his book. This was a direct revelation of God's words. "Is it such a fast I have chosen? a day for a man to afflict his soul? Is it to bow down his head like a bulrush, and to spread sackcloth and ashes under him? Wilt thou call this a fast, and an acceptable day to the Lord? Is not this fast I have chosen: to loose the bands of wickedness, to undo the heavy burdens, and to let the oppressed go free, and that ye break every yoke? Is it not to deal thy bread to the hungry, and that thou bring the poor that are cast out into thy house? when thou seest the naked, that thou cover him; and that thou hide not thyself from thine own flesh? Then shall thy light break forth as the morning, and thy health shall spring forth speedily: and thy righteousness shall go before thee; the glory of the Lord shall be thy rereward. Then shalt thou call, and the Lord will answer; thou shalt cry, and he shall say, Here I am", (Isaiah 58: 5–9). Jesus said, "Let your light so shine before men, that they may see your good works, and glorify your Father which is in heaven" (Matthew 5:16).

Tantum Ergo

Jesus said to the Samaritans,
"You worship you know not what".
He wasn't referring to Sinai,
Or to the burning bush;
Not to the shekinah glory
That at one time filled the tent;
Not to lofty Mount Gerazim,

Or to Old Jerusalem sent.

He meant that forms of worship
Did not of themselves suffice
However attractive to proselytes
Who'd forgotten the golden calf.
God is no longer looking for sacrifice,
But desires obedience instead.
So what is the value of ritual
If there's no difference within?

Worship offered by sweeping up,
Doing kindnesses for neighbours,
Showing others by practical help
God's reality by loving labours.
It is not by bowing and scraping
We necessarily show a proof:
It is by lives demonstrating how
To worship in spirit and truth.

Dr. V. John Delany, from an impending Anthology

Bibliography

Alexander, Achibald. *Thoughts on Religious Experience.* Banner of Truth. Third Edition, 1967.
Abbott, Walter. S.J. *The Documents of Vatican II.* Geoffrey Chapman. London, 1966.
Bauckham, Richard. *The Climax of Prophecy.* T. & T. Clark. Edinburgh, 1993.
Bright, J. *A History of Israel.* SCM Press, 1967.
Bromley, David G. *Falling from the Faith.* Sage Publications, 1988.
Caird, G.B. *The Revelation of St. John the Divine.* A. & C. Black. London, 1984.
Davies, J.M. *The Lord and the Churches.* Pickering and Inglis, 1967
Delany, Dr. V. John. *Here I Stand.* Centenary Address. *Churchman,* Vol. 143, 1989.
Delany, Bernard. O.P., B. Litt. (Oxon.). *The True History of Maria Monk.* Catholic Truth Society, 1943.
Farrar, F. W. *The Early Days of Christianity.* Cassell & Co., London, 1884.
Hendricksen. *More Than Conquerors.* InterVarsity Press, 1962.
Hick, John. *The Myth of God Incarnate.* SCM Press, 1997.
Hislop, Rev. Alexander. *The Two Babylons.* S. W. Partridge & Co., Paternoster Row, London, 1871.
Hodge, A.A. *The Westminster Confession.* Banner of Truth, 1998.
Lang, G.H. *Gospel of the Kingdom.* G.H. Lang, 1957.
Lang, G.H. *First Fruits and Harvest.* G.H. Lang, 1947.
Ladd, Eldon. *Jesus and the Kingdom.* SPCK, 1966.
Luther, Martin. *Commentary on St. Paul's Epistle to the Galatians.* First published 1535. James Clarke & Co., Ltd, 1953.
Maritain, Jacques. St. Thomas Aquinas. *Angel of the Schools.* Trans. J. F. Scanlan. Catholic Book Club. London, 1947.
More, Thomas. *Utopia. Translated into Modern English.* By G. C. Richards. Oxford Basil Blackwell, 1923.
Newman, J.H., Cardinal. *Apologia Pro Vita Sua.* J. M. Dent & Sons Ltd, 1955.
Owen, John. *The Works of John Owen,* Vol. 4. Banner of Truth, 1974.
Pinnock, Clark J. *Most Moved Mover.* Paternoster Press, 1937.
Pinnock, Clark J. *Towards a New Theology.* Creation House, 2001.
Pope John Paul II. *Veritatis Splendor.* Papal Encyclical Letter.
Selwyn, E.G. *The Teaching of Christ.* Longman Green & Co, 1915.
Sheehan, M. – Archbishop. *Apologetics and Catholic Doctrine.* M. H. Gill & Sons Ltd., Dublin, 1950.
Spector, R. H. *Eagle Against the Sun.* Viking, 1985.

Thomas, L.R. *A Symposium of Prophecy.* Albany, Western Australia, 1947.
Wagner, Peter. *Territorial Spirits.* Sovereign World, 1991.
Wells, David. A. F. *God in the Wasteland.* Inter Varsity Press, 1973.
Wells, H. G. *Crux Ansata.* Penguin, 1943.
Wojtyla, Karol. *Through the Year with Pope John Paul* II. Ed. Tony Castle. Crossroads. New York, 1981.
Yeats, W. B. *Memoirs.* Macmillan, 1974.

About the Author

The author is a former Director in the Civil Service and Past President of a chartered professional body. He holds a Ph.D. from the University of Loughborough in Human and Environmental Studies and was awarded an honorary fellowship by Bath Spa University for his outstanding contribution to the development of British higher education. He has publications in this field, and numerous published papers (some in the Christian field). He is also an Associate of The Michaelis Guild of The Froebel Institute (University of London at Roehampton) and former Governor. Involvement in Christian work includes association with The Scripture Gift Mission, Christian Service Training Council, and The Open Theological College. He has been honorary lecturer in Contemporary Religion at Redcliffe Missionary Training College, visiting preacher across several denominations, and an overseas consultant on three continents. After active service in the Royal Navy in World War II he married and has a son, a grandson, and two great granddaughters. He is a member of Frinton Free Church and has been a member of Gideons International for many years.

Bible Reference Index

Versions drawn on for bible references are based mainly on either The Authorised or New International Bibles. Version is not considered important under the heading "reference" as the sense is the same. There is one reference to the J. B. Philips modern translation.

Preface

PAGE	Reference	Keyword
viii	Matthew 3:2	Repent
viii	Matthew 10:37	Worthy
ix	John 5:28	Graves
ix	1 Corinthians 14:8	Trumpet
ix	Colossians 1:13	Darkness
x	Ephesians 2:12	Hope

Introduction

PAGE	Reference	Keyword
1	Matthew 13:31	Mustard
1	Matthew 13:3	Sower
1	Acts 1:3	Rose
1	Revelation 1:9	Patmos
1	Matthew 16:18	Prevail
2	Mark 7:26	Syrophenician
2	Matthew 8:5	Centurion
2	Luke 10:33	Samaritan
2	Ezra 5:12	Nebuchadnezzar
3	Matthew 11:17	Piped
3	Luke 9:54	Fire
3	1 Corinthians 6:11	Some of you
3	1 Corinthians 4:6 / John 21:25 21:25	Written
3	1 Corinthians 13:12	Glass darkly
4	2 Peter 3:16	Understand
4	2 Peter 1:3	Godliness
5	Galatians 5:17	Conflict
6	2 Peter 1:16 J.B. Philips	Cleverly

| 6 | Romans 10:20 | Found |
| 6 | Matthew 6:33 | Seek |

Part One   God's Present Kingdom

Chapter 1

PAGE	Reference	Keyword
11	Romans 14:17	Joy
12	John 16:8	Judgment
12	Galatians 5:19-21	Witchcraft
12	Galatians 5:22-23	Fruit
12	Revelation 7:9	Multitude
12	1 Corinthians 2:9	'No eye has seen'
13	2 Corinthians 12:4	Unutterable
13	Ezekiel 1:5	Ezekiel
13	Revelation 21:1	John
13	John 18:36	'Not of this world'
14	Matthew 21:31	Publicans and harlots
14	Matthew 23:13	Prevents /shut
14	John 3:19	Darkness
14	Romans 10:3	Ignorant
14	Matthew 13:47	Net
14	Luke 17:20	Observation
14	Matthew 20:10	Penny
14	Ephesians 3:8	Unsearchable
15	Matthew 13:25	Weeds
15	Matthew 16:18	'My Church'
15	John 16:8	Judgement

Chapter 2

PAGE	Reference	Keyword
17	Genesis 1:29, 30	Carnivores
17	2 Corinthians 3:18	Likeness
17	Genesis 12:1	Country
17	Hebrews 11:10	Architect
17	Hebrews 11:7	Righteousness
17	The Book of Judges	Amphictyony
17	Ezra 5:12	Nebuchadnezzar
17	John 18:36	Of this world
17	Romans 9:6	Israel
18	Matthew 27:51	Curtain
18	Romans 10:12	Jew
18	Luke 10:18	Lightning
18	Matthew 11:12	Violently
18	Luke 2:25	Simeon

18	2 Corinthians 4:18	Eternal
19	Acts 2:1	Pentecost
19	Acts 2:42	Fellowship
19	Matthew 13:31	Mustard
19	Ephesians 5:27	Wrinkle
19	Revelations 3:17	Rich
20	2 Peter 1:4	Partake
20	Daniel 7:15	Daniel
21	Matthew 3:16	Spirit
21	Hebrews 12:28	Shaken

Chapter 3

PAGE	**Reference**	**Keyword**
22	2 Corinthians 4:3	Hidden
23	Matthew 7:14	Narrow
23	Luke 18:25	Needle
23	Luke 15:11	Prodigal
24	2 Corinthians 4:4	God
25	1 Peter 1:4	Away
26	John 10:1	Sheepfold
26	Matthew 13:3	Sower
26	Matthew 7:21	Lord
27	Philippians 1:11	Righteousness
27	Ephesians 5:27	Present
28	Matthew 24:36	Time
28	2 Peter 3:12	Fire
28	1 Corinthians 13:12	Glass
28	Philippians 3:15	Differently
28	Mark 9:38	Independently
28	Matthew 5:6	Hungering
29	Luke 1:38	Willing
29	Luke 1:38	Handmaid
29	1 Timothy 2:5	Mediator
29	1 Peter 5:8	Roaring
29	James 4:7	Resist
29	2 Peter 2:5	Noah
29	Genesis 6:4	Giants
30	Matthew 6:13	Evil
30	Luke 11:20	Finger
30	Matthew 12:29	House
30	Luke 18:42	Sight
30	2 Corinthians 4:4	Glorious

BIBLE REFERENCE INDEX | 177

Chapter 4

PAGE	Reference	Keyword
32	Romans 14:17	Joy
32	1 Corinthians 14:33	Confusion
33	1 Corinthians 14:36	Word
34	Mark 10:28	Everything
34	1 Corinthians 12:9	Gift
34	1 Corinthians 15:17	Raised
34	James 5:16	Another
34	Genesis 3:17	Adam
34	Luke 4:27	Syrian
34	Matthew 12:40	Whale
34	John 3:5	Birth
35	2 Timothy 1:6	Stir
35	Romans 1:18	Unrighteousness
35	Matthew 5:13	Salt
35	2 Timothy 1:7	Timidity
35	Matthew 5:14	Hill
35	Malachi 3:16	Spoke

Chapter 5

PAGE	Reference	Keyword
38	Colossians 1:13	Delivered
38	1 Peter 2:9	Marvellous
38	1 Peter 1:4	Inheritance
38	Matthew 16:19	Keys
38	Acts 2:38	Repent
38	1 Peter 1:3	Lively
39	1 Peter 3:21	Dirt
39	Acts 9:3	Damascus
39	John 6:44	Draws
39	Romans 10:13	Calling
39	Romans 10:9	Confessing
39	Ephesians 4:15	Growing
39	Matthew 5:6	Hungering
39	Romans 8:16	Witness
39	Matthew 16:18	Prevail
39	Romans 8:39	Separate
40	2 Peter 2:5	Ark/Noah
40	Acts 17:30	Winked
40	Ephesians 2:1	Transgression
40	John 1:13	Flesh
40	1 Peter 3:18	Unjust
40	Romans 1:4	Power
40	John 12:24	Corn

Chapter 6

PAGE	Reference	Keyword
41	Matthew 21:28	Sons
41	Philippians 1:11	Fruits
41	2 Corinthians 5:14	Constrain
41	Romans 3:24	Justified
41	Romans 10:3	Ignorant
41	James 2:20	Faith
42	Ephesians 2:8	Grace
42	Ephesians 2:10	Works
42	Luke 9:62	Plough
42	Matthew 7:17	Fruit
42	Revelations 3:19	Chastened
42	Psalm 73:17	Sanctuary
42	Galatians 5:17	Conflict
43	Luke 9:23	Cross
43	Romans 6:6	Crucified
43	Romans 6:11	Reckoning
43	Philippians 3:3	Confidence
43	Romans 6:4	Newness
43	1 John 5:18	Begotten
43	1 John 2:1	Advocate
43	Ephesians 1:20	Exerted
43	Ephesians 4:11	Ministry
43	Ephesians 6:11	Armour
43	Matthew 13:46	Pearl
43	Colossians 1:26	Ages
43	2 Corinthians 4:7	Earthen
44	1 John 3:14	Death
44	Romans 5:17	Reign
44	Colossians 3:3	Hidden
44	1 Corinthians 12:27	Body
44	James 2:16	Warm
44	Matthew 25:43	Strangers

Chapter 7

PAGE	Reference	Keyword
46	Hebrews 11:6	Rewarder
46	Genesis 15:1	Reward
46	Daniel 9:24	Seventy
47	John 1:46	Nazareth
47	Matthew 2:7	Herod's
47	Micah 5:2	Bethlehem
47	Luke 24:12	Tomb
47	Acts 12:13	Peter

BIBLE REFERENCE INDEX

47	2 Timothy 1:12	Believed
47	John 9:25	Blind
47	Exodus 23:11	Fallow
49	Acts 18:3	Tent
49	1 Timothy 5:18	Ox
50	Luke 12:27	Lilies
50	Luke 7:6	Mustard
50	Matthew 8:8	Centurion
50	Matthew 8:11	Abraham
50	Ephesians 2:10	Foreordained
51	1 Corinthians 12.9	Gift
51	Mark 11:22	Faith
51	Matthew 24:13	End
51	Revelation 20:5	First
51	Revelation 20:5	Second
51	Romans 8:18	Compared

Chapter 8

PAGE	**Reference**	**Keyword**
54	Hebrews 12:1	Cloud
54	Hebrews 12:2	Fix
54	Hebrews 12:28	Moved
55	Hebrews 12:12	Feeble
55	Romans 13:1	Authority
55	2 Corinthians 10:4	Carnal
55	Acts 5:29	Obey
55	1 Peter 4:1	Suffered
55	Matthew 26:29	Vine
57	Acts 17:23	Unknown
57	Galatians 5:22	Fruit

Chapter 9

PAGE	**Reference**	**Keyword**
59	Genesis 1:3	Light
59	Hebrews 4:12	Sharper
59	2 Corinthians 3:6	Kills
59	John 16:14	Conveys
59	John 15:1	Vine
60	Romans 10:14	Heard?
61	Matthew 24:35	Pass

Chapter 10

PAGE	**Reference**	**Keyword**
63	John 1:14	Flesh
63	Revelation 1:5	First

PAGE	Reference	Keyword
63	Matthew 4:3	Stones
64	Luke 4:18	Proclaim
64	John 16:7	Expedient
64	John 16:14	Glorify
64	John 15:5	Vine
64	Matthew 28:18	Authority
64	John 1:33	Baptiser
64	Acts 2:16	Joel
65	2 Corinthians 3:18	Likeness
65	Colossians 1:27	Mystery
65	1 Timothy 2:5	Mediator
65	John 16:8	Righteousness
65	Ephesians 1:20	Exercised
65	1 Corinthians 12:7	Profit
65	Psalm 51:11	Spirit
66	Ezekiel 10:18	Temple
66	Romans 8:26	Intercedes
67	1 Thessalonians 5:20	Despised
67	1 Corinthians 14:5	Interpretation
68	Acts 2:16	Joel
68	1 Thessalonians 5:19	Quench
68	2 Peter 2:1	False
68	2 Corinthians 4:7	Treasure
68	Galatians :11	Withstood
68	1 Corinthians 2:4	Enticing
69	Isaiah 59:19	Flood

Chapter 11

PAGE	Reference	Keyword
70	Genesis 3:4	Serpent's
70	Galatians 1:6	Different
71	1 Thessalonians 5:23	Spirit
71	James 3:15	Earthly
72	Hebrews 2:14	Destroy
72	2 Timothy 3:5	Denies
73	John 11:43	Lazarus
74	1 Samuel 28:7	Endor
74	Numbers 3:4	Strange fire
74	Acts 19:14	Sceva
74	Colossians 2:15	Principalities

Chapter 13

PAGE	Reference	Keyword
80	Ephesians 5:17	Unwise
83	Acts 16:16	Divination

85	2 Corinthians 10:	Imaginations
85	Galatians 4:8	Nature
85	Galatians 4:9	weak
85	Galatians 4:17	Zealous
85	1 Corinthians 1:18	Foolishness
85	1 Corinthians 1:19	Destroy
85	1 Corinthians 1:20	Foolishness

Chapter 14

PAGE	**Reference**	**Keyword**
86	Acts 12:7	Angel

Chapter 15

PAGE	**Reference**	**Keyword**
92	Matthew 6:9	Prayer
92	Galatians 4:6	Abba
93	Matthew 6:7	Repetitive
93	Numbers 11:9	Manna
93	John 6:35	Bread
93	Romans 10:9	Confess
93	Matthew 18:22	Seventy
94	John 6:35	Life
94	Luke 24:49	Tarry
94	1 Timothy 5:5	Supplication
94	John 17:4	Completed
94	Colossians 1:27	Glory
94	James 5:14	Elders
95	Matthew 6:5	Lord's
95	Romans 8:26	Groans
95	Matthew 6:5	Street

Chapter 16

PAGE	**Reference**	**Keyword**
97	Luke 14:26	Hate
97	2 Peter 3:3	Scoffers
97	1 Thessalonians 5:2	Thief
98	Genesis 8:21	Aroma
99	Genesis 19:7	Women
99	Hebrews 13:2	Entertaining
99	John 7:19	Moses
101	Colossians 1:27	Glory
102	Galatians 3:24	Schoolmaster
103	Galatians 5:11	Offence
103	John 15:18	Hate

Part Two   The Coming Kingdom

Chapter 17

PAGE	Reference	Keyword
108	Luke 4:18	Ministry
108	Luke 24:13	Emmaus
108	Revelation 19:11	Horses
108	Revelation 8:6	Trumpets
108	Revelation 4:7	Eagles
109	2 Peter 2:1	Teachers
109	1 John 4:1	False
110	Romans 9:6	Israel
110	Hebrews 4:14	High Priest
110	Hebrews 7:17	Melchizedek
111	Revelation 19:16	Kings
111	Zechariah 14:4	Zechariah
111	John 16:11	Prince
112	2 Corinthians 4:16	Inwardly
112	2 Timothy 4:8	Attained
112	Matthew 7:3	Speck

Chapter 18

PAGE	Reference	Keyword
114	Revelation 6:10	Long
115	Acts 2:41	Thousands
115	Daniel 2:33	Clay
116	1 John 4:3	Anti-Christ
117	Luke 23:28	Weep
117	Matthew 24:36	Know
117	Matthew 16:28	Taste
117	Acts 1:3	Kingdom
117	Luke 6:24	Seven
117	Acts 9:3	Damascus
120	Matthew 5:5	Meek
120	Luke 22:30	Twelve
120	Matthew 19:29	Houses
120	Luke 22:29	Conferred
120	Matthew 24:40	Two
120	1 Corinthians 15:52	Twinkle
120	Revelation 7:1	Corners
120	Revelation 14:18	Fruits
120	Revelation 14:15	Grain
120	Hebrews 9:27	Appointed
121	Matthew 25:1	Virgins
121	Matthew 25:15	Talents

PAGE	Reference	Keyword
121	Matthew 25:32	Goats
121	Luke 12:46	Pieces
121	Mark 13:26	Clouds
121	Acts 1:11	Stand

Chapter 19

PAGE	Reference	Keyword
124	Revelation Ch2&3	Letters
128	Revelation 20:5	1,000 years
128	Revelation 20:27	Lamb's
128	Revelation 2:10	Death
128	Revelation 11:3	Witnesses
129	Matthew 24:40	Two
129	Revelation 19:20	Lake
129	Romans 8:22	Groaning
129	Isaiah 11:6	Lamb
129	Revelation 15:3	Moses
129	Exodus 1:16	Boy
129	Revelation 20:4	Reign
129	Revelation 14:14	Sickles
129	Revelation 14:18	Grape
129	Revelation 20:11	Throne
129	Hebrews 9:27	Appointed
130	1 Corinthians 3:12	Hay
130	Matthew 7:23	Depart
130	Revelation 20:12	Works
131	2 Corinthians 5:1	Celestial
131	John 13:5	Washing
132	Revelation 16:9	Curse
132	2 Corinthians 7:10	Sorrow
133	Revelation 22:11	Vile
133	Galatians 6:7	Sows
133	Revelation 11:13	Repentance

Chapter 20

PAGE	Reference	Keyword
136	Revelation 18:4	Come out
136	Revelation 18:10	Hour
137	1 Peter 2:1	Malice
137	2 Chronicles 32:10	Sennacherib
138	Revelation 16:16	Armageddon
138	Revelation 19:20	Beast
139	Matthew 5:6	Thirst

Chapter 21

PAGE	Reference	Keyword
140	Revelation 13:17	Mark
141	Exodus 13:21	Pillar
141	Hebrews 4:6	Unbelief
141	Revelation 15:3	Moses
141	Matthew 26:64	Mighty
142	Revelation 20:4	Reign
142	John 1:51	Angels
143	Romans 8:18,20	Frustration
143	Revelation 22:2	Healing
143	Revelation 22:1	Flowing
144	2 Peter 3:10	Day
144	Isaiah 24:20	Drunkard
144	2 Peter 1:19	Prophecy
144	Numbers 24:17	Jacob
144	Revelation 22:16	Root
144	Revelation 5:5	Judah
144	Isaiah 60:3	Brightness
144	Zechariah 9:9	Colt
144	John 12:24	Grain
144	1 Thessalonians 4:13	Asleep
145	1 Corinthians 15:54	Incorruption
145	2 Corinthians 5:4	Swallowed
145	John 20:27	Thomas
145	Revelation 20:15	Lamb's
146	Revelation 20:10	Fire
146	Revelation 19:20	Prophet
146	1 John 3:2	Like
146	1 Corinthians 2:10	Reveal
147	Isaiah 32:1	Righteousness
147	Isaiah 35:8	Holiness

Chapter 22

PAGE	Reference	Keyword
148	Genesis 3:8	Walking
148	Genesis 3:8	Hid
148	Genesis 5:24	Enoch
148	Genesis 12:1	Abraham
148	Exodus 3:2	Bush
148	Exodus 19:20	Sinai
149	Hebrews 12:23	Heaven
149	John 1:14	Grace
149	1 Corinthians 12:13	Jews
149	Revelation 21:5	Behold

149	Exodus 40:34	Glory
149	Revelation 22:5	Lights
149	Isaiah 60:61	Arise
149	Acts 9:3	Damascus
150	Isaiah 62:4	Hephzibah
151	Matthew 24:35	Pass
151	Revelation 21:3	Dwelling
151	Revelation 21:2	Bride
151	Revelation 19:7	Marriage
151	Matthew 26:29	Vine
151	1 Corinthians 15:47	Earthy
181	Galatians 5:17	Contrary
152	Revelation 19:16	Kings
152	Revelation 19:13	Splattered
153	Isaiah 9:6	Wonderful
153	Hosea 7:8	Ephraim

Part Three Summary and Conclusions

Chapter 23

PAGE	**Reference**	**Keyword**
158	John 1:51	Descending
158	Luke 24:25	Prophets
158	Psalm 96:10	Reigns
158	Revelation 20:7	Release
158	Isaiah 11:9	Waters
159	Revelation 14:9	Mark
159	Philippians 3:11	Resurrection
159	Revelation 20:5	Raised
159	Revelation 14:15	Reaping
159	Deuteronomy 11:14	Grain
159	Revelation 14:18	Grape
159	Luke 3:17	Garner
159	Revelation 14:20	Trodden
159	Revelation 21:26	Come in
159	Revelation 21:3	Tabernacle
159	Revelation 21:3	People
161	Psalm 2:4	Laugh
161	Genesis 1:31	Good
161	Matthew 3:13	Jordan
161	Matthew 10:7	At hand
161	1 Timothy 3:16	Vindicated
161	Romans 8:29 / Revelation 1:5	First born

161	Acts 1:3	Kingdom
163	Luke 9:62	Plough
164	2 Timothy 2:17	Hymenaeus
164	2 Timothy 1:15	Phygelus
164	2 Timothy 1:15	Hermeogenes
164	John 17:24	World
164	Matthew 7:29	Authority
164	John 8:58	Abraham
164	Mark 13:26	Clouds
167	Matthew 5:3	Poor
168	Matthew 5:6	Thirst
168	Matthew 22:26	Important
168	Matthew 5:5	Meek
168	John 4:21	Worship
168	1 Corinthians 15:24	Hand Over
168	Hebrews 8:11	Neighbour

Subject Index

Africa, 48, 49, 53, 83, 88, 89
AIDS, 29, 108, 137
Arminianism, 26

Babylon, 4, 22
Babel, 13
British Israel, 108, 109
Brunei, 14

Calvinism, 24, 26
CERN Collider, 138
Charismatics, 68
Chilianism, 162
Counterfeit, 83, 84

Da Vinci Code, 6, 24, 164
Defection, 30
Denominationalism, 27, 100, 101

Eschatology, Alternative, 107; Olivet Discourse, 114; Revelation, 124; Babylon, 134; Christ's Kingdom, 140; New Jerusalem, 148

False Prophets, 109

Gallicans, 27, 33, 101
Gideons, 6, 44

Horoscopes, 74, 75–79

Inquisition, 118

Jesuits, 33

Kingdom of God, Nature, 11; History, 16; Hidden, 22, 32; Entering, 38; Life, 41; Faith, 46; Servants, 53; Work, 58; Holy Spirit, 63; Evil, 70; Horoscope, 75; Wisdom, 80; Testimonies, 86; Prayer, 92

Media, 36
Medicine Sans Frontiere, 44
Millennialism, 44, 110, 111, 131, 136, 143
Missionary Aviation Fellowship, 45
Muslims, 53, 57, 137

Nicolaitans, 125

Openness, 82
Opus Dei, 62
Overcomers, 125
Oxford Movement, 102

Papal Encyclicals, Dignitatis Humanae, 167; Gaudium et Spes, 167; Humanum Genus, 136; Quadragesimo Anno, 56; Rerum, Novarum, 56; Veritatis Splendor, 166, 167
People, Allenbury, General, 119; Aquinas, Thomas, 27; Augustine, of Hippo, 33; Aylward, Gladys, 54; Bauckham, Richard 140; Benson, Hugh, 73; Beveridge, William, 56; Booth, William, 169; Byron, Lord, 137; Calvin, John, 135; Catherwood, Lady, 90; Carey, William, 56; Carmichael, Amy, 54, 56; Challoner, Bishop, 169; Chesterton, G.K., 33; Churchill, Winston, 81; Cohen,

Gordon, 164; Damien, Father, 169; Dollinger, Arch Bishop, 169; Elwes, Valentine, 90; Fletcher, John, 55; Gandhi, Mahatma, 56, 81; Graham, Billy, 67; Hinderer, Anna, 54; Hislop, Alexander, 135; Hume, Basil, 101; Kaunda, Kenneth, 57, 89; Kempis, Thomas, 169; King, Martin, Luther, 56; Kuhn, Isobel, 54; Ligouri, Alphonsus, 33; Lincoln, Abraham, 28; Lloyd-Jones, D.M., xi, 90; Loyola, Ignatius, 33; Lupino, Father, 169; Luther, Martin, 33, 42, 160, 169; Mandela, Nelson, 56, 89; Montgomery, Bernard, 24; More, Thomas, 11; Muller, George, 25, 51; Newman, J.H., 25, 33, 61, 101, 118, 135; O'Connell, Patrick, 90; Owen, John, 67, 109, 120; Paisley, Ian, 165; Reagan, Ronald, 136; Romaine, William, 49; Roosevelt, Franklin, 23, 24, 166, 167; Rossler, Otto, 138; Ryle, Bishop J.C, 169; Serra, Father, 169; Schleirmacher, Frederick, 169; Sheeham, Archbishop, 119; Slessor, Marie, 54; Studd, C.T., 50; Sutherland, "Mammy", 54; Taylor, Hudson, 51, 58, 169; Tyndale, William, 169; Townsend, Henry, 54; Vickers, Nora, 90; Wesley, Charles, 42; Wesley, John, 55; Whitefield, George, 55; Wilberforce, William, 169; Wycliffe, John, 169; Yeats, W.B., 78

Pilgrim Fathers, 132

Popes, Benedict XVI, 27, 119; Celestine, 119; Gregory VII, 162; John Paul II, 165, 166; Leo XIII, 56, 62, 136; Pius XI, 56

Protestantism, 102, 118, 168

Revivals, 65, 66

Roman Catholicism, 4, 22, 27, 29, 61, 118

Roman Empire, 4

Roman Emperors, Antiochus Epiphanies, 122; Celestine Gallus, 122; Nero, 134; Constantine, 135, 162

Séances, 73

Scripture Gift Mission, 50

Society of St. Vincent de Paul, 45

Suicides, 72

Testimonies, 86, 87, 88

Territorial Spirits, 81

Tradition, 41

Ultramontanes, 101, 168

Utopia, 11, 161

Vatican II, 27, 62, 101

Virgin Mary, 27, 29

Witchcraft, 73, 84